Creep and Shrinkage of Concrete Elements and Structures

Developments in Civil Engineering

Developments in Civil Engineering, 21

Creep and Shrinkage of Concrete Elements and Structures

by

ZDENĚK ŠMERDA

Research Institute of Engineering Structures
Brno, Czechoslovakia

and

VLADIMÍR KŘÍSTEK

Czech Technical University
Prague, Czechoslovakia

ELSEVIER

AMSTERDAM – OXFORD – NEW YORK – TOKYO 1988

This book is the updated translation of
Dotvarování a smršťování betonových prvků a konstrukcí
published by SNTL, Prague, 1978

Translated by Ing. Prokop Maxa
Translation Editor: J. J. Brooks, University of Leeds
Published in co-edition with SNTL — Publishers of Technical Literature, Prague

Distribution of this book is being handled by the following team of publishers
    for the U.S.A. and Canada
Elsevier Science Publishing Company, Inc.
52 Vanderbilt Avenue
New York, N.Y. 10017

    for the East European Countries, China, Northern Korea, Cuba, Vietnam and Mongolia
SNTL — Publishers of Technical Literature
Spálená 51, 113 02 Prague

    for all remaining areas
Elsevier Science Publishers
25 Sara Burgerhartstraat
P.O. Box 211, 1000 AE Amsterdam, The Netherlands

Library of Congress Cataloging-in-Publication Data

Šmerda, Zdeněk.
    Creep and shrinkage of concrete elements and structures.
(Developments in Civil Engineering; 21)
    Translation of: Dotvarování a smršťování betonových prvků a konstrukcí.
    Includes bibliographies and index.
    1. Concrete construction — Testing. 2. Concrete — Creep. 3. Concrete — Expansion and contraction.
I. Křístek, Vladimír. II.Title.
TA681.S5713 1988   620.1'3633   87-27159

ISBN 0-444-98937-4 (Vol. 21)
ISBN 0-444-41715-X (Series)

Printed in Czechoslovakia

Contents

5

Preface

The creep and shrinkage of concrete are complex phenomena which are not yet understood completely. Creep and shrinkage have a considerable impact upon the performance of concrete structures, causing deflection increases as well as affecting stress distribution. In analysing the effects of creep and shrinkage of concrete we not only come across the problem of safety against failure, but we also find ourselves dealing with very important economic factors such as durability, serviceability and long-time reliability. This is why there is now a pressing need to solve the fundamental problems of creep and shrinkage and to develop adequate methods of predicting the structural performance of concrete members and structures affected by them.

The aim of this book is to discuss the results of experimental research, to summarize various approaches to the prediction of creep and shrinkage, and to present suitable methods of structural creep analysis which allow the creep and shrinkage effects to be simply calculated. Particular attention is paid to methods which are intended for use as design tools.

The first chapter of the book deals with the fundamental features of creep and shrinkage of concrete according to the results of experimental research. The second chapter is devoted to the mathematical modelling of creep and shrinkage of concrete and to the practical prediction of these phenomena. Several practical models for predicting mean cross-section creep and shrinkage are also presented in this chapter.

Chapter 3 deals with the analysis of concrete members at the cross-sectional level with particular attention directed to the non-homogeneity of cross-sections due to different creep properties and the influence of reinforcement. The fourth chapter presents methods of creep and shrinkage analysis of concrete structures. The chapter covers a wide range of problems for various types of structures (e.g. the effects of changes in the structural system, the buckling of compression members under long-term loads, the load distribution in cellular structures affected by creep, etc.).

Chapter 5 deals with the structural performance of concrete members and structures in which cracks may develop. The sixth chapter has a special intention — practical experience with effects of creep and shrinkage is discussed here and some efficient methods to reduce these effects are proposed. This represents an active approach to structural design which tries not only to explain and quantify the creep and shrinkage effects, but also presents recommendations on how to minimize their negative influence while at the design stage. This is also the main intention of the book as a whole.

It is hoped that the present book will encourage progress in further research and simultaneously provide a useful basis for practical structural design.

Zdeněk Šmerda
Vladimír Křístek

1
Experimental findings and discussion on creep and shrinkage

1.1 Shrinkage and creep — their interrelationship, and relaxation

The basic characteristic of the strains of concrete, involving shrinkage, swelling and creep is that their development and magnitude are dependent on time. Shrinkage and swelling develop without any action of stresses, since they are caused solely by water passing from concrete into the environment and vice versa. On the other hand, creep depends on the stress of concrete as well as on water movement.

1.1.1 Strains

Let us first, at least partly, differentiate between strain which is independent of time and strain which is time-dependent.

In principle, every strain is a function of time because a stress is always introduced into the concrete during a definite time interval (even a very small one), and therefore the distinction between a time-dependent strain and a strain not depending on time is very uncertain. However, to be more specific, let us refer to tests on concrete specimens. Here, it is assumed that the strain is measured when the specified stress has been reached. Strains obtained in this way shall be considered to be independent of time, i.e. "instantaneous".

Then the time-dependent, or "delayed", strains are measured from the time, when the instantaneous strains have developed.

Both instantaneous and delayed strains are reversible or irreversible (permanent strains). The reversible strains may be divided into instantaneous elastic (ε_e) and delayed elastic ($\varepsilon_{e,d}$) strains; the irreversible strains into instantaneous inelastic strains (ε_{ne}) of a plastic nature, and delayed inelastic strains ($\varepsilon_{ne,d}$) of a viscous nature. If all these strains are considered, including those independent of load, i.e. strains caused by temperature and humidity and by shrinkage ε_s, they may be tabulated as shown in Table 1.1.

Table 1.1   Classification of strains

Strain	dependent on stress		independent of stress
	instantaneous	delayed	
reversible	elastic ε_e	elastic $\varepsilon_{e,d}$	caused by temperature and humidity
irreversible	inelastic ε_{ne}	inelastic $\varepsilon_{ne,d}$	caused by shrinkage or swelling ε_s

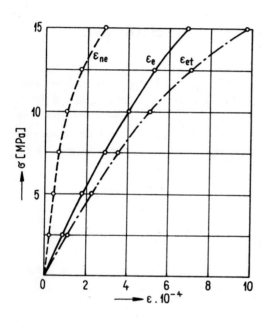

Fig. 1.1 Instantaneous strain of concrete

This classification of strains is employed in Ref. [1.1], introduced there from [1.2].

Now, according to the foregoing definitions, the creep strain ε_c should be a time-dependent strain which must equal the sum of the delayed elastic strain and of the delayed inelastic strain, viz.

$$\varepsilon_c = \varepsilon_{e,d} + \varepsilon_{ne,d} \tag{1.1}$$

Instantaneous strains

The total instantaneous strain of concrete ($\varepsilon_{e,t}$) always consists of the elastic ε_e (reversible) and inelastic ε_{ne} (irreversible) components; their distribution is shown in Fig. 1.1 where it may be observed that

12

(a) the permanent strain (ε_{ne}) is small at a low stress level in concrete and increases non-linearly with the increasing stress;

(b) the elastic strain (ε_e) also is non-linearly dependent on stress. However, its distribution is nearly linear under lower stresses. Hence, the assumption of linearity is admissible.

Instantaneous elastic strain

For concrete structures, the allowable stress of concrete equals only about a half of its strength, and the ratio σ/ε_e is considered constant up to the value of allowable stress, so that the instantaneous elastic strain is

$$\varepsilon_e = \sigma/E_c \tag{1.2}$$

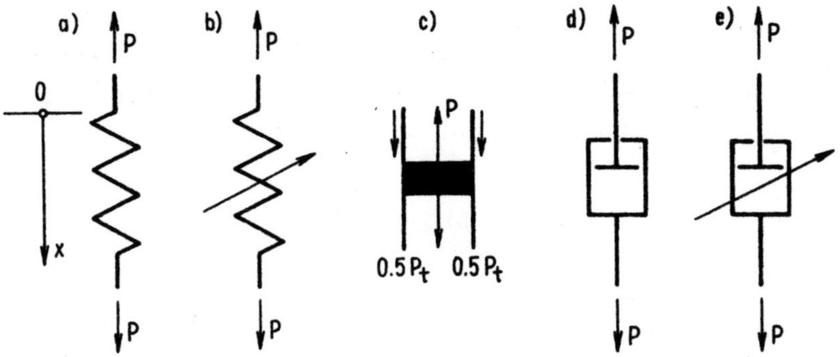

Fig. 1.2 The elements of rheological models

A material satisfying this relationship would have to be perfectly linearly elastic and can be modelled by a helical spring (Fig. 1.2a) whose extension x is directly proportional to the force P. Hence

$$x = \bar{\alpha}P \tag{1.2a}$$

Note: The elastic properties of concrete depend on its age because the modulus of elasticity of concrete $E_c(t)$ changes in time. Therefore, Eq. (1.2) should be

$$\varepsilon_e(t) = \sigma/E_c(t) \tag{1.3}$$

while the relationship for the model should also be

$$x(t) = \bar{\alpha}_t P \tag{1.3a}$$

In model rheology, an elastic material with time variation $\bar{\alpha}_t$ is represented by a helical spring with an arrow (Fig. 1.2b); neither the stress nor the force P in Eqs. (1.2) and (1.3) are functions of time.

13

Instantaneous inelastic strain

The instantaneous irreversible strain ε_{ne} is explained by means of the plasticity of concrete. Concrete has no clear limit of plasticity, which is really the condition of plastic flow (this occurs only when the stress σ just exceeds the limit stress σ_T, i.e., the yield strength). However, as the instantaneous inelastic strain is a permanent strain attained in a very short time interval, it may be explained merely by the dislocation of the structural elements of the material, a process which corresponds to the plastic manifestation.

Plastic strain may be modelled by a piston frictionally connected with the walls of the cylinder (Fig. 1.2c). As long as the force P is smaller than the resistance P_T generated by the friction between the piston and the cylinder walls, no displacement (x) of the piston occurs. Only when P just exceeds the value of the force P_T does the piston start to move, as long as the force P is applied. If the force P is discontinued, the piston does not return into its original position. Hence, the relationship between the plastic strain ε_p and time is such that, from the instant when $t = 0$ and the stress has reached the value σ_T at which a displacement occurs, this strain increases proportionally with time (Fig. 1.3a).

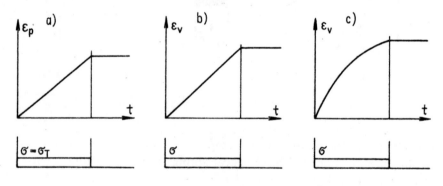

Fig. 1.3 Plastic and viscous strains

Delayed strains

Delayed inelastic strains ($\varepsilon_{ne,d}$) are explained by means of the viscous property of concrete ($\varepsilon_{ne,d} = \varepsilon_v$), when water is extruded from the micropores of the gel under the effect of stress. In this process, the strain velocity depends on the magnitude of stress σ and on the viscosity of the material:

$$\frac{d\varepsilon_v}{dt} = v\sigma \tag{1.4}$$

where ε_v designates viscous strain.

14

The model corresponding to this ideal viscous material (when $v_0 =$ constant) is represented by a piston exerting pressure on a liquid with a viscosity v_0 (Fig. 1.2d). Owing to the effect of force P, the liquid escapes around the piston, which moves as a consequence. When the force ceases to act, the motion of the piston is discontinued.

The displacement x is given as follows:

$$v_0 \frac{dx}{dt} = P \qquad\qquad (1.4a)$$

Hence,

$$x = \frac{1}{v_0} Pt, \quad \text{if} \quad x = 0 \quad \text{for} \quad t = 0,$$

and

$$\varepsilon_v = v\sigma t \qquad \text{(Fig. 1.3b)}.$$

To explain the strain of concrete caused by viscosity, it is necessary, in cases of more complex stresses, to consider viscosity $v(t)$ variable in time. The relation (1.4a) is then modified to

$$\frac{d\varepsilon_v}{dt} = v(t)\,\sigma \qquad\qquad (1.4b)$$

The model is now represented by a cylinder and a piston with an arrow (Fig. 1.2e) and the relationship between the strain and time may develop, for example, in accordance with Fig. 1.3c.

Delayed elastic strain $\varepsilon_{e,d}$ may be imagined as a common behaviour of elastic and viscous materials. Its rheological model is portrayed in Fig. 1.4a (Kelvin's model), where the elastic and viscous elements are linked in parallel. As both Eqs. (1.2a) and (1.4a) apply:

$$P = \frac{x}{\bar{\alpha}} + v_0 \frac{dx}{dt} \qquad\qquad (1.5)$$

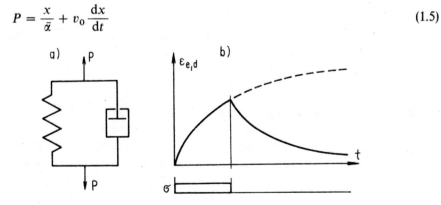

Fig. 1.4 Elastic delayed strain

while the displacement under the initial condition, when $x = 0$ for $t = 0$, is

$$x = P\bar{a}(1 - e^{-\frac{t}{\bar{a}v_0}}) \tag{1.6}$$

where e is the base of natural logarithms.

The displacement x corresponds to the strain of concrete $\varepsilon_{e,d}$, hence the distributions of both quantities are similar; the relationship between $\varepsilon_{e,d}$ and t is shown in Fig. 1.4b. Evidently, the process reverses at unloading, the strain $\varepsilon_{e,d} \to 0$ for $t \to \infty$ and the strain is fully reversible.

Theoretically, it would be possible to explain all the components of the total strain $\varepsilon_t = \varepsilon_e + \varepsilon_{ne} + \varepsilon_{ne,d} + \varepsilon_{e,d}$ with reference to the foregoing discussion. In spite of this, the interpretation of the individual strains by different workers, particularly $\varepsilon_{e,d}$ and $\varepsilon_{ne,d}$, differs considerably.

Note: All rheological models are idealized because they assume materials with ideal properties. In reality, for instance, an increase of the modulus of elasticity of concrete with age affects the constant $\bar{a}$ with the result that certain strains remain in the concrete even after the unloading of the "elastic" model.

Strains obtained from test results

Many tests have been carried out with the purpose of finding strain of concrete and its division into the elastic component ε_e, creep ε_c and shrinkage ε_s (Table 1.1), and some of the results are presented here.

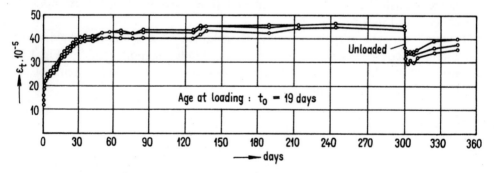

Fig. 1.5 Total strain of concrete prisms 100/100/400 mm; strength = 35.30 MPa

Figure 1.5 shows the test results reported in [1.18]. The specimens were subjected to a load at the age of 19 days (t_0 is the age of concrete at the commencement of loading), after which they were under a stress $\sigma_0 = 4.45$ MPa for 300 days and then they were unloaded. The process of loading lasted about 5 minutes; as the stress σ_0 was low, the plastic instantaneous strains $\varepsilon_{ne} = \varepsilon_p$ could be neglected. Consequently, the curves express the distribution and the magnitude of the total strain, i.e. of the elastic strain ε_e, creep ε_c and shrinkage ε_s. The strains ε_c and ε_s are characterized by

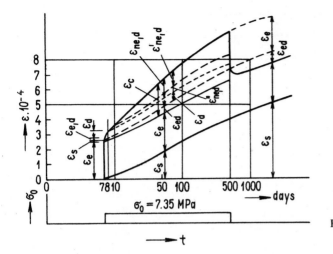

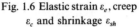

Fig. 1.6 Elastic strain ε_e, creep
ε_c and shrinkage ε_{sh}

a high rate of growth in the initial stage after the loading and by a subsequent asymptotical approach to their final value. When the specimens were unloaded, the strain representing shrinkage of concrete was still increasing.

The distribution of the individual strains is shown in Fig. 1.6, which illustrates the tests reported in [1.4] where the specimens were loaded for 7–500 days. On analysing the strain, the authors arrived at the conclusion that creep ε_c contained not only the components $\varepsilon_{e,d}$ and $\varepsilon_{ne,d}$ but that, in a brief interval following the loading (approximately in one day), there developed a certain strain ε_d (rapid initial strain) which had an irreversible character. Probably, this again is a plastic strain which is formally separated from the instantaneous irreversible strain ε_{ne}, because the development of the latter is conditioned by a time interval of longer duration. Hence, creep becomes

$$\varepsilon_c = \varepsilon_d + \varepsilon_{e,d} + \varepsilon_{ne,d}$$

The problem, however, is still not fully clarified. In the tests, it appeared that the viscous part of the strain $\varepsilon_{ne,d}$ was partly reversible, which led to its division into basic creep ($\varepsilon'_{ne,d}$) and drying creep ($\varepsilon''_{ne,d}$). The difference in the two types of creep is supposed to express the opinion that the "basic creep" develops in concrete when no water passes between the concrete and the environment (tests on sealed specimens) and that "drying creep" is that which develops in the concrete besides the basic creep. The creep $\varepsilon''_{ne,d}$ is partly reversible, partly permanent, $\varepsilon'_{ne,d}$ is irreversible.

Evidently, the problem of creep is intricate from the point of view of phenomenology, and the more so from a fundamental viewpoint. Although the magnitudes of the individual strains have been investigated, no reliable information was obtained as to the parts which should be attributed to plastic and viscous changes, respectively. It appears, however, that the share of plastic strains in creep is less

17

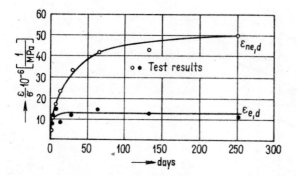

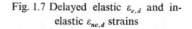

Fig. 1.7 Delayed elastic $\varepsilon_{e,d}$ and inelastic $\varepsilon_{ne,d}$ strains

substantial; consequently, the explanation of creep on the basis of elastic delayed and inelastic delayed strains will be adhered to in the discussions. The tests of the magnitudes of each of these strains yield more reliable results, and their dependence on time has been satisfactorily investigated. The velocities of the individual strains and their magnitudes, portrayed in Fig. 1.7, are in agreement with the tests presented in [1.7]. It may be concluded that the total strain is

$$\varepsilon_t = \varepsilon_e + \varepsilon_{e,d} + \varepsilon_{ne,d} + \varepsilon_s \tag{1.7}$$

in which the inelastic instantaneous strain (ε_{ne}) is considered negligible and the rapid initial strain ε_d included in the strain $\varepsilon_{ne,d}$.

It should be emphasized that all discussions concern a state in which concrete is subjected to comparatively small stresses in relation to its strength. In the opposite case, not only the strains resulting from the plastic deformation of concrete, but also those due to the formation of microcracks, become significant.

In the case of the elastic strain ε_e, the variation of the modulus of elasticity in time is usually not taken into account, because the errors involved are smaller than those arising from the calculation of long-term strains (creep and shrinkage) calculated by means of formulae with large tolerances. The variation of the modulus of elasticity in time (i.e. its increase) is considered only in more accurate analyses, and the elastic strain then also depends on time; its development under a constant stress is shown in Fig. 1.8a, together with the actual distribution and magnitude of

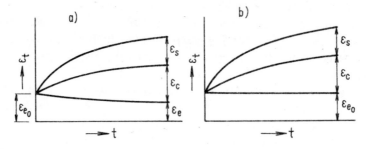

Fig. 1.8 Long-term total strain

the total strain ε_t. If E_c is independent of time, the strains develop in accordance with Fig. 1.8b. Evidently, the total strain is somewhat larger with E_c = constant.

With reference to Eq. (1.7), it is assumed that all the strains can be added, i.e., the individual strains are not interdependent. Although it has been proved by tests that this is not the case in the relationship between creep and shrinkage, such a simplification is admissible, and its practical consequences for the analysis are very important.

1.1.2   Relaxation

If a concrete structural member can freely deform under a permanent constant stress, its deformation increases due to creep. If the free development of deformation from creep is prevented, the original stress is reduced, i.e., relaxation takes place. Consequently, the relaxation of stress is interrelated with creep. In the analyses of relaxation, the findings concerning the development and magnitude of creep are

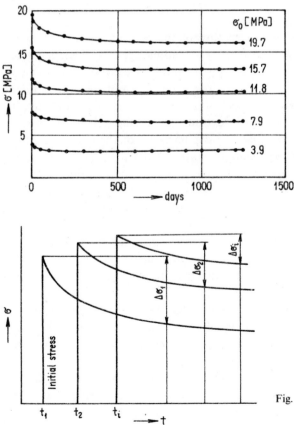

Fig. 1.9 Effect of the magnitude of the initial stress σ_0 on the relaxation process

Fig. 1.10 Decrease of stress under constant original strain

usually employed, because the measurements of creep do not present the difficulties inherent in the measurements of relaxation, and the available tests of stress relaxation confirm the interrelation between relaxation and creep.

The development of the relaxation process at different levels of the initial stress is plotted in Fig. 1.9; it may be observed that the relaxation process develops more rapidly (compared with creep) at the beginning and approaches its final value asymptotically. The loss of stress $\Delta\sigma$ in concrete specimens subjected to a stress at different ages of concrete (t_1, t_2, t_i), is shown in Fig. 1.10, where it can be seen that the stress decreases at a higher rate in a younger concrete. The differences in relaxation of the initial stresses are caused here by the different values of the modulus of elasticity of the concrete, which is variable in time $(E_{ct_2} > E_{ct_1})$, because the initial strains of all the specimens were equal.

1.2 Experimental findings on creep and shrinkage

1.2.1 Creep

The origin of creep is in the microstructure of the cement paste binding the aggregate and the sand grains. The basis of this binding agent is the cement gel, which is a very homogeneous material with a colloidal character; it contains chemically bonded water, colloidal water in the gel pores and free water in the capillaries and macropores. Under the effect of a long-term stress in concrete, the water, which is not bonded chemically, is extruded from the gel micropores (of an order one hundred times smaller than the capillaries) into the capillaries, from which it evaporates. While the extrusion of water is determined by the stress of concrete, the evaporation depends on the hygrometric conditions of the ambiance. Owing to the loss of water, the stress is transferred gradually from the viscous medium onto the elastic skeleton of the concrete; after unloading, the strain reappears to some extent.

Hence, the magnitude of creep depends on the stress in concrete, on its structure with regard to the properties of its individual components, on the consistency of the mix and on the degree of hydration. It is affected also by the ambient conditions and temperature.

Three methods are mostly employed to express creep: either the strain (ε_c) directly, or the "specific creep" $C = \varepsilon_c/\sigma$ is employed, the latter being the creep due to a unit stress; the third method employs the creep coefficient, $\varphi = \varepsilon_c/\varepsilon_e$, which is the ratio of creep and elastic strain.

When the time-dependent distribution and the final value of creep are known, its magnitude at any instant t may be found, the assumption being that linear relationship of stress versus strain and the law of superposition hold true.

1.2.1.1 *Linearity and creep*

In the linear theory of elasticity, stresses are directly proportional to strains. One of the fundamental problems in the investigation of creep is the experimental proof that such linear relationship exists between the stress and creep at every instant t, i.e.

$$\varepsilon_c(t) = A(t)\,\sigma \qquad (1.8)$$

Here, $A(t)$ depends on time and on the modulus of elasticity $E_c(t)$ of concrete. If this is true, then the relationship

$$\varepsilon_c(t) = \frac{\varphi(t)}{E_c(t)}\,\sigma = \varphi(t)\cdot\varepsilon_e(t)$$

also holds true and, with a constant modulus of elasticity of concrete, we may write

$$\varepsilon_c(t) = \varphi(t)\,\sigma/E_c \qquad (1.9)$$

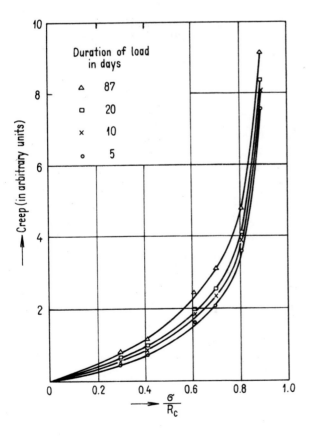

Fig. 1.11 The relation between creep and σ/R_c

21

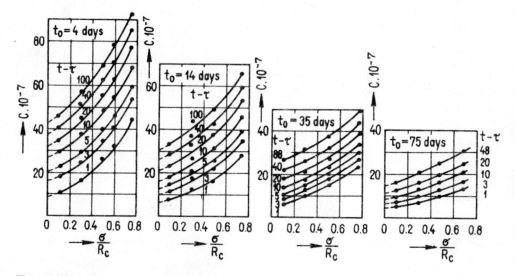

Fig. 1.12 Effect of the ratio of stress σ to the prismatic strength R_c of concrete on creep in compression in the interval $\langle t - \tau_0 \rangle$

where $\varphi(t)$ is the creep coefficient representing the ratio of creep ε_c at a definite instant and the elastic strain ε_e, i.e.

$$\varphi(t) = \varepsilon_c(t)/\varepsilon_e \tag{1.10}$$

Many tests, aiming to confirm the relation (1.8) under uniaxial compression, have indicated that the assumption of linearity is acceptable in practice but that linearity applies only up to a certain limit of the ratio of the stress σ to the strength of concrete R_c; when this limit is exceeded, creep ε_c is no longer proportional to the stress any more. The test results introduced in [1.6] (Fig. 1.11) show that creep accelerates significantly around 0.7 of the ratio σ/R_c; hence, this might represent the limit of this ratio. It may also be deduced from the curves that the limit of σ/R_c depends on the duration of the stress, because the abrupt change of gradient of the curve is not apparent for a concrete loaded for 87 days.

The tests carried out in [1.9] (Fig. 1.12), when the specimens were loaded at time t_0, lead to the conclusion that there exists no clearly defined limit of the ratio σ/R_c. In spite of this, it is proposed to assume a linear relationship at the values of the ratio σ/R_c not exceeding 0.5; prismatic strength is used for the strength of concrete. A lower limit results from other tests; for example, in [1.8], the range of 30–40 per cent of the ratio σ/R_c is applicable.

It may be concluded that the linearity of stress versus creep is justifiable for stresses not exceeding 40–50 per cent of the strength of concrete. In practical applications, this is acceptable, because, under service conditions, the stress from the principal load varies around 35 per cent of the strength of concrete in compression or the strength in bending. Only in special instances, such as in the case of a local stress caused by concentrated load, will the ratio be higher.

The validity of the principle of linearity in the case of uniaxial tension has not been sufficiently investigated until now. Some of the available results, however, indicate that the strain versus stress relationship is linear even when the ratio of tensile stress to tensile strength of concrete approaches unity [1.9].

The principle of linearity has not been confirmed experimentally even for a multiaxial state of stress; however, in the analyses of such cases, proportionality is assumed. Under this assumption, the creep coefficient $\varphi(t)$ should be the same for a concrete subjected to compression or tension, the creep Poisson's ratio $v_1(t)$ should be the same as the elastic Poisson's ratio $v(t)$ and, finally, the relationships between the modulus of elasticity of concrete under compression $E_{cc}(t)$ and under tension $E_{ct}(t)$ and the modulus of elasticity in shear $G_c(t)$ should be the same as if the concrete behaved elastically. If $E_{cc}(t) \doteq E_{c_t}(t) = E_c(t)$; $v_1(t) = v(t)$, then

$$G_c(t) = \frac{E_c(t)}{2[1 + v(t)]} \tag{1.11}$$

where the average value of the elastic Poisson's ratio $v(t)$ is 0.167. The results of research for the creep Poisson's ratio $v_1(t)$ given in the literature, for example [1.10, 1.11, 1.12], suggest that $v_1(t)$ may have a much higher value than is the elastic value. On the other hand, some workers found lower values: for example, in accordance with [1.14], the value of $v_1(t)$ was within the range of 0.16 to 0.25, while, according to [1.15], the range is between 0.17 and 0.20.

Bearing in mind these conflicting data, it is difficult to decide on a definite value of $v_1(t)$, although it seems that the values of both the Poisson's ratios $v_1(t)$ and $v(t)$ could be approximately the same. An average value of 0.167 is recommended rather than using possibly extreme values of $v_1(t)$ which could lead to error in the analysis of a combined state of stress. It has been shown by numerical comparisons [1.13] that the differences do not exceed 5–6 per cent; consequently, the assumed equality of $v(t)$ and $v_1(t)$ is justified.

1.2.1.2 *Superposition*

The superposition of strains results from the assumption of a linear relationship between stress and strain. Evidently it also applies when stresses of different values are introduced into the concrete at the same time τ_0; the resulting creep ε_c at time t equals the sum of increments of creep ε_{ci} under a stress σ_i. Hence

$$\varepsilon_c(t) = \Sigma \varepsilon_{ci} = \frac{\varphi(t)}{E_c} \Sigma \sigma_i$$

For the analysis of creep, the validity of superposition is generalized for the cases when concrete is subjected to stresses at different times. If superposition

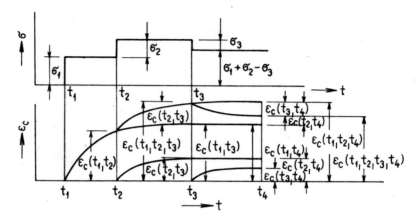

Fig. 1.13 Superposition of creep

holds good, then (in accordance with Fig. 1.13), creep $\varepsilon_c(t_1, t_2, t_3)$, i.e. the creep at time t_3 from the stress σ_1 acting within the interval $\langle t_1, t_3 \rangle$, and from the stress σ_2 acting within the interval $\langle t_2, t_3 \rangle$, is as follows:

$$\varepsilon_c(t_1, t_2, t_3) = \varepsilon_c(t_1, t_3) + \varepsilon_c(t_2, t_3)$$

where

$\varepsilon_c(t_1, t_3)$ is the creep from the stress σ_1 acting within the interval $\langle t_1, t_3 \rangle$;
$\varepsilon_c(t_2, t_3)$ is the creep from the stress σ_2 acting within the interval $\langle t_2, t_3 \rangle$.

If a part of the stress is removed, this part may be considered negative and creep ε_c is subtracted. Hence, if the stress is reduced by σ_3 within a time interval $\langle t_3, t_4 \rangle$ (Fig. 1.13), and the respective creep is $\varepsilon_c(t_3, t_4)$, then the total creep at time t_4 is

$$\varepsilon_c(t_1, t_2, t_3, t_4) = \varepsilon_c(t_1, t_4) + \varepsilon_c(t_2, t_4) - \varepsilon_c(t_3, t_4) =$$
$$= \varepsilon_c(t_1, t_2, t_4) - \varepsilon_c(t_3, t_4)$$

It is evident from the principle of superposition that the individual stresses always act during the entire time interval, i.e. from the instant when the stress has been introduced into the concrete until the observed time.

The experiments have confirmed a satisfactory agreement between the test results and the values calculated under the assumption of the validity of superposition, but only when the stress increased. When it decreases, superposition is applicable only to a limited extent when the decrease of stress is not large. If the drop of stress is large, the actual decrease of creep is less than that calculated.

The results of tests carried out on members subject to bending [1.16] are compared with the calculated values of creep in Fig. 1.14, which shows that, when the member is unloaded suddenly and completely, the permanent strain is somewhat higher than the calculated one. Another comparison [1.20], Fig. 1.15

24

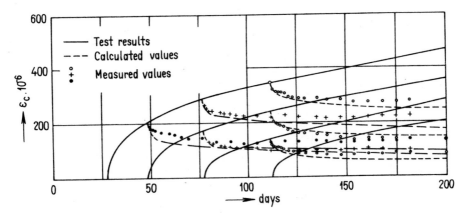

Fig. 1.14 Strains after unloading of concrete

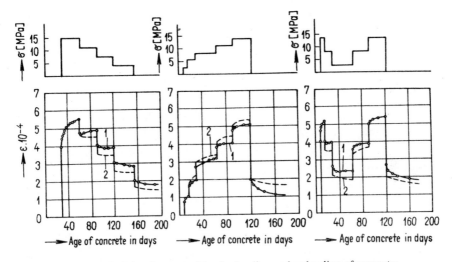

Fig. 1.15 Principle of superposition by loading and unloading of concrete:
1 — test results; 2 — results obtained from the solution based on the principle of superposition

shows that, even under a gradually increasing stress, the actual creep is less than the calculated one; here, the test results are shown by solid lines, while the results calculated in accordance with the principle of superposition are shown by dashed lines. In all instances, the numerical analysis was based on experimentally found distributions and magnitudes of creep, and the discussed differences might engender a criticism on the principle of superposition and creep. However, the quoted experiments demonstrate that the differences are not too large, especially in the case of stresses encountered in practice. Hence, the principle of superposition is satisfied to a large extent and, in fact, has been generally accepted, because, without it, a practical analysis of creep would be impossible.

1.2.1.3 *Factors influencing creep*

A number of factors affecting the magnitude of creep has been mentioned in the discussion; each of them influences creep to a different degree or magnitude. Consequently, only the factors with the highest significance are introduced into the analysis.

All methods of estimating creep consider the effect of the ambient conditions (storage) and of the age of concrete at loading, other methods contain detailed instructions making it possible to express the effect of the dimensions of the structural members, the consistency of concrete, the kind of cement and aggregate. The latter factor has acquired significance especially recently, because of an increase in the uses of artificial porous aggregate and structural lightweight concretes.

The effect on creep of the individual factors has been studied in many research institutes and the results have been published in many papers. While some factors have been treated thoroughly, evaluation of other factors is only just beginning. Here, we present the results of research on factors already incorporated in the Codes and Standards, or on factors which should be considered in the assessment of creep.

The individual factors are interrelated, and they affect not only the final magnitude of creep, but also its development, i.e., its value at every instant t. For this reason, creep is frequently expressed with the aid of coefficients which sometimes depend on the same factors.

Thus, for example, in the Soviet literature [1.18] the following expression is introduced for the analysis of creep:

$$\varepsilon_c = \frac{\sigma}{E_c} \varphi_t = \frac{\sigma}{E_c} \varphi_\infty (1 - e^{-Bt}) = \frac{\sigma}{E_c} (1 - e^{-Bt}) \varphi_\infty^c \eta_1 \eta_2 \eta_3$$

where the term $1 - e^{-Bt}$ expresses the time-dependent distribution of creep (B being a constant) and φ_∞ designates the final value of the creep coefficient. This coefficient consists of four components: φ_∞^c expresses the effect of the water/cement ratio, composition of concrete, admixtures, type of aggregate and of the mixing method; η_1 expresses effect of the ambient conditions; η_2 allows for the dimensions of the structure; η_3 allows for the age of concrete. Similar coefficients are employed in the CEB-FIP Recommendations 1970 (Sect. 2.6.1). In view of their interrelation, the individual factors cannot be separated, and this should be borne in mind when they are analysed.

Age of concrete

It has been explained earlier that the development of creep in concrete under stress is rapid at its onset but slows down subsequently. According to some sources, 80–90 per cent of the total creep develops in the course of the first year, and creep is almost terminated by the end of five years. Measurements carried out on actual structures, however, have indicated that creep occurred even after 20 years.

Creep in concrete subjected to stress not exceeding approximately 0.4 of its strength develops in agreement with the curve shown in Fig. 1.16 [1.21], the shaded area showing the scatter of the measured values. In the present case, the test specimens were loaded at the age of 14 days and, consequently, the growth of creep is very rapid in the first stage.

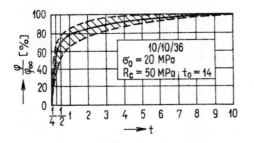

Fig. 1.16 Creep at constant load, and constant temperature and humidity; R_c is the cube strength of concrete

In accordance with the experimental results, the later concrete is loaded, the lower will be the final value of creep; also, the time-dependent distribution will not be the same. This finding is of primary significance not only for the determination of the final value of creep but also for the determination of the relations expressing its dependence on time. Figures 1.17 and 1.18 show the results of the measurements of creep [1.18, 1.19] on laboratory specimens. In either instance, the concrete was loaded at a different age and the effect of the age of concrete at loading (instant t_i) on the magnitude of creep at an arbitrary instant t is evident. For different ages, t_i, of concrete at loading, the distribution diagrams of creep have different characters, because (Figs. 1.17 and 1.18) the angles α_1, $\alpha_2, \ldots \alpha_5$ of the tangents to the curves, corresponding to ages of concrete t_1, $t_2, \ldots t_5$ are not equal.

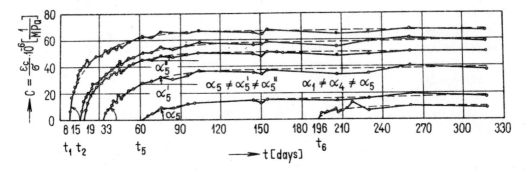

Fig. 1.17 Creep related to the age t_0 at loading

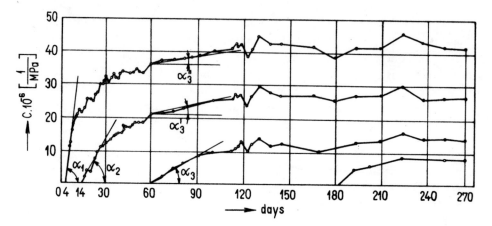

Fig. 1.18 Creep of concrete prisms loaded at the age of 4, 14, 60 and 81 days

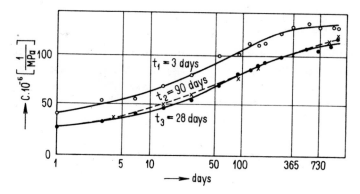

Fig. 1.19 Effect of the age t_1, t_2, t_3 at loading on creep

The tests depicted in Figs. 1.17 and 1.18 also show that the age $t_i = t_0$ at which the concrete was loaded strongly influences the final creep value as $t \to \infty$; referring to Fig. 1.17, the specific creep C is 38 for concrete loaded at the age $t_0 = 33$ days, while for $t_0 = 196$ days, C is reduced to 9. Other tests (for example, shown in Fig. 1.19 [1.22]) indicate that, if $t_0 > 28$ days, the age of concrete at loading does not have any significant effect either on the distribution or on the final value of creep.

The quoted results disagree to some extent because the foregoing conclusion could apply to $t_0 \doteq 90$ days and not to $t_0 = 28$ days, according to the tests shown in Figs. 1.17 and 1.18. This disagreement is evidently of a quantitative character, since, in principle, there must be a relationship between the two curves for the entire range of t_0, in accordance with Fig. 1.20 [1.21], and furthermore, these creep diagrams prove that there is a large effect of the age at loading for $t_0 > 28$ days.

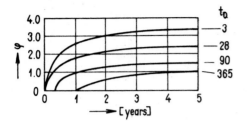

Fig. 1.20 Creep coefficient φ of concrete loaded at different ages

All values presented in the graphs are relative to the actual magnitudes of creep; creep is expressed either by its specific value by the creep coefficient, or by the ratios of both these terms. To obtain the actual values, it is necessary to consider more factors, such as the influence of the ambient conditions, in particular, the dimensions of the structural members, the consistency of concrete, and the kind of cement.

The test results of several authors [1.23] are compared in Fig. 1.21 to obtain an objective assessment of the effect of age of concrete. The values of the coefficient k_d, expressing the influence of concrete age at instant t_0, when stress was introduced into the concrete (see Sect. 2.6.1), are plotted as the ordinate, while the time, t_0, measured from the age of casting the test specimen, is plotted as the abscissa. The results, when arranged in this way, have a visible descending trend with the growing age t_0. This tendency may be portrayed by a theoretical curve (curve A), for which two analytical expressions have been derived [1.23]:

$$k_d = 0.36 + \frac{26.3}{13 + t_0} \quad \text{for the entire range of } t_0;$$

$$k_d = 0.36 + 1.15\,e^{-0.021 t_0}, \quad \text{when } t_0 > 28 \text{ days.}$$

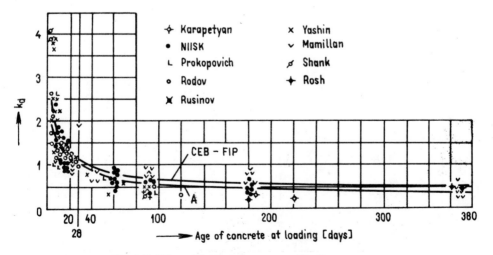

Fig. 1.21 Effect of the age of concrete at loading on creep

The other curve in the graph has been calculated from the relations in accordance with the Recommendations CEB-FIP (see Sect. 2.6.2).

The distribution and magnitude of creep with relation to the age of concrete at loading time t_0 and with relation to an arbitrary instant is discussed in Sect. 2.1.

Ambient conditions (storage)

The final values of creep depend on hygrometric conditions: under drier conditions creep is large, while in water it is small or none. According to tests [1.24], the final value of creep depends linearly on relative humidity (Fig. 1.22); for a duration of load of 14 days, creep is similar over the relative humidity range of 50 to approximately 75 per cent. Direct proportionality between creep and relative humidity is therefore assumed for test duration of 200 days and beyond.

The effect of the ambient conditions on the magnitude and time-dependent distribution of creep is depicted in Fig. 1.23 [1.29]; concrete specimens were normally cured for 28 days, then they were loaded at the age $t_0 = 28$ days and stored at 50, 70 and 100 per cent relative humidities. The development of creep at 100 per cent relative humidity was different from creep under drying conditions.

A difference in the time-dependent creep distribution of concretes loaded at different ages is noticeable in the tests depicted in Fig. 1.24 [1.21], where specimens

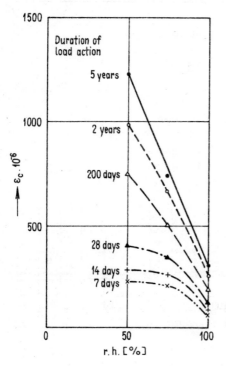

Fig. 1.22 The relation between relative humidity and creep; dimensions of specimens: 70/70/280 mm and stress = 10.0 MPa

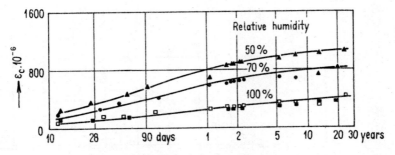

Fig. 1.23 Effect of relative humidity on creep

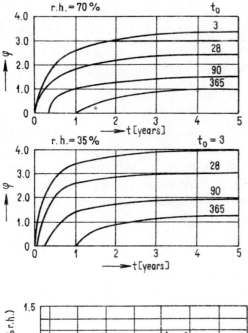

Fig. 1.24 Effect of the age at loading on creep with regard to humidity

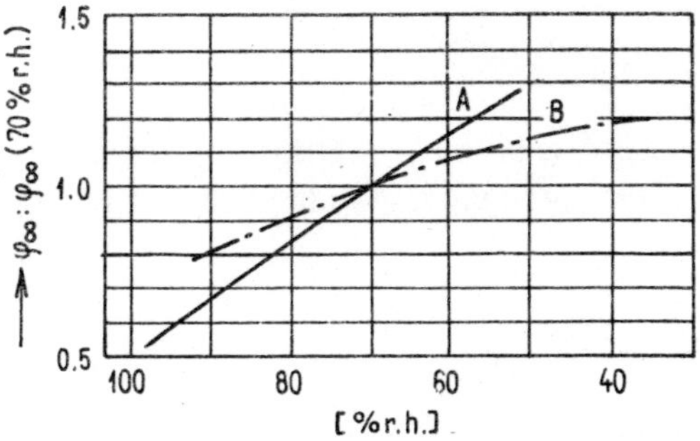

Fig. 1.25 Effect of type of cement on creep

were stored at 70 to 35 per cent relative humidities. Under drying conditions, creep develops at a higher rate in the initial period after loading than under more humid conditions.

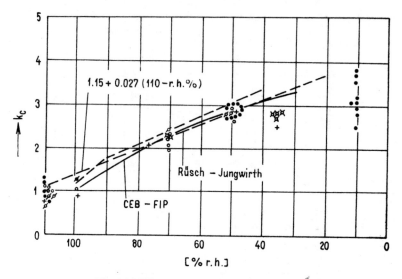

Fig. 1.26 Effect of relative humidity on creep

The behaviour depends on many factors which can have varying effects on the results, which explains the differences in the test results as presented by individual authors. Figure 1.25, for example, illustrates the effect of different types of cement on the creep coefficient (φ) stored at different relative humidities [1.8]. The creep coefficient ratio $\varphi_\infty/\varphi_{\infty(70)}$ is shown, when φ_∞ is the final creep coefficient at any humidity, and $\varphi_{\infty(70)}$ is the final magnitude of the creep coefficient at 70 per cent relative humidity; curve *A* is related to normal-hardening cement and curve *B* to rapid-hardening cement.

The test results of several authors are summarized in Fig. 1.26 in terms of the coefficient k_c which expresses the effect of relative humidity on creep. The quoted linear relationship

$$k_c = 1.15 + 0.027 \, (110 - \text{per cent r.h.})$$

satisfactorily expresses the experimental results; a bi-linear relationship represents the proposal referred to in [1.4] and the solid curve corresponds to the CEB–FIP Recommendations [1.1].

Cross-sectional dimensions of structural member

Structural members of smaller dimensions creep more than large members. The important factor is the area of the surface which is exposed to the environment and is instrumental in transferring moisture from the concrete to the environment. The distribution and magnitude of creep of structural members of various sizes are shown in Figs. 1.27 and 1.28. The difference in creep of structural members

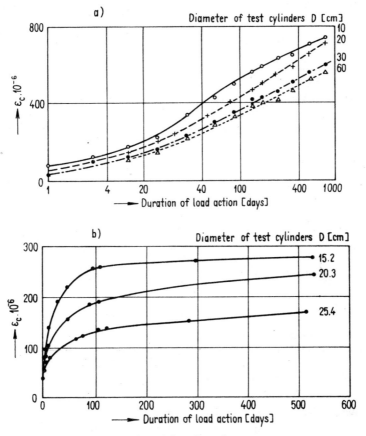

Fig. 1.27 Effect of size of member on creep

with various sizes may be explained by a more intensive exchange of moisture in the case of smaller members; hence, only the so-called "drying creep" $\varepsilon''_{ne,d}$ is considered.

The test results of several authors, who investigated the relationship between creep and the dimensions of the structural members, are portrayed in Fig. 1.28. The coefficient k_e, representing the effect of the dimensions on creep (Sect. 2.6.1), is plotted on the ordinate, and the smallest dimension d_{min} of the member is plotted on the abscissa. The solid line (up to $d_{min} = 120$ cm) represents a curve introduced in [1.23], where creep is related only to the smallest dimension; however, this curve does not express reality quite correctly. Other recommendations [1.1] consider the so-called effective thickness of the structural member, which is expressed by the area of the cross-section divided by half its circumference, i.e.

$$d_f = \frac{2A}{O}$$

33

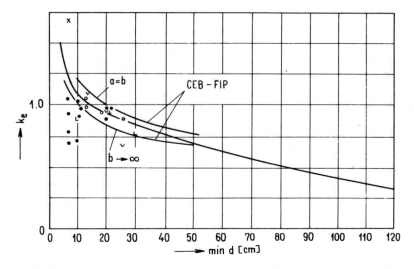

Fig. 1.28 Relation between the coefficient k_e and the smallest dimension of a member

where A is the area and O is the circumference of the cross-section. This expression is more satisfactory, as it takes into account both dimensions.

Creep determined in accordance with the CEB–FIP

Recommendations cannot be compared with the test results in [1.23], which is evaluated in relation to the effective thickness d_f in the first case, and by means of the thickness d_{min} in the second case. Consequently, the comparisons in Fig. 1.28 are given for two cases: one concerns a structural member with a square cross-section $(a = b)$ and the other applies for a plate-shaped member, where $b \rightarrow \infty$. In the first case, $d_f = \frac{1}{2}a$, while in the second case, $d_f \rightarrow a = d_{min}$. Evidently, the relationships obtained by the CEB–FIP reliably cover the zones of the average values obtained from the test results; the only exception are very thin members, where, however, the results are more scattered.

Type of cement

In the understanding and assessment of creep, the properties of the cement gel have a decisive significance, and these properties are directly related to the properties of cement. It has been emphasized that the principal agent of creep is the cement paste, while the skeleton consisting of aggregate creeps little.

The relationship between creep and the type of cement is complex. It depends on the physical and chemical properties of the cement, and also on the degree of

34

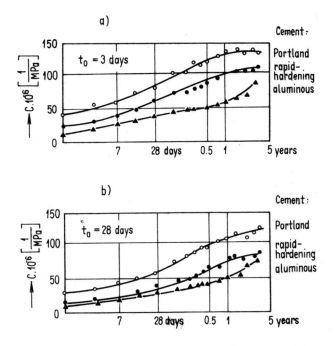

Fig. 1.29 Effect of type of cement on creep (t_0 is the age at loading)

hydration of the cement paste. Here, only the differences appearing in the distributions and magnitudes of specific creep will be discussed for three types of cement: Portland, rapid-hardening and aluminous. The results of measurements are presented in Fig. 1.29 [1.26]. In the first case, specific creep is related to members loaded at the age $t_0 = 3$ days, and in the second case, the members were subjected to a load at the age of 28 days. The different creep values are undoubtedly affected by the fact that concretes prepared of different cements attain the same strength at different times, a process which also strongly affects the final creep value. However, the very steep growth of creep after one year in a concrete made from aluminous cement is very surprising, because it disagrees with the idea of an asymptotic distribution of creep. Since the tests confirm the significance of the type of cement on creep, some Codes express this relationship with the aid of empiric coefficients modifying the final value of creep (Sect. 2.3).

Type of aggregate

It has been proved by research that the type of natural aggregate has an effect on the magnitude of creep, but this is not taken into account in the formulae for the analysis of creep. For the time being, this approach is justified, because the effect of the type of aggregate is still not known sufficiently so that it can be

35

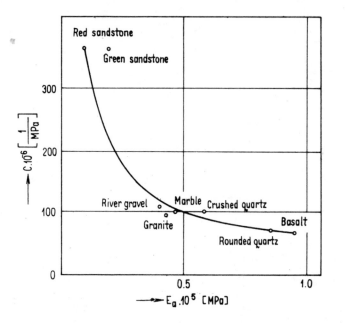

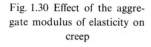

Fig. 1.30 Effect of the aggregate modulus of elasticity on creep

introduced into the analysis, and also because, in most cases, the designer of a structure does not know in advance which type of aggregate will be used for the preparation of concrete.

The problem changes, however, with artificial porous aggregate whose use for the manufacture of structural concretes has increased in recent years. For these cases, all Codes account for the differences in the magnitude and distribution of creep between concretes made with natural dense aggregates and artificial porous aggregates.

Tests indicate that creep depends on the modulus of elasticity E_a of the aggregate, which is much lower for porous artificial aggregate than for a natural aggregate. Thus, for example, E_a is approximately 0.045×10^6 MPa for granite, 0.04×10^6 MPa for pebble and 0.012×10^6 MPa for artificial porous aggregate. Such differences cannot be neglected in the assessment of creep. Research has been undertaken [1.28] to establish the relationship between the modulus of elasticity of aggregate and creep; this relationship for some natural aggregates is shown in Fig. 1.30 in terms of specific creep C. The distribution and final magnitudes of creep of concrete prepared from some types of aggregate are presented in Fig. 1.31 [1.29].

Creep of concretes prepared from lightweight aggregate is also not known to the extent that reliable information is available about its distribution and magnitude; one of the reasons is that concretes prepared from various artificial porous aggregates creep differently. Creep of a lightweight concrete may be lower or higher than that of a normal concrete of the same strength, as shown in Fig. 1.32; here, the zone indicates the range of specific creep C for concretes of different

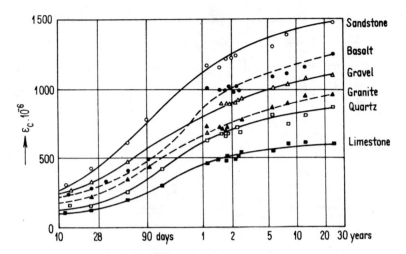

Fig. 1.31 Creep of concretes made from different aggregates; specimens subjected to a stress of 5.6 MPa

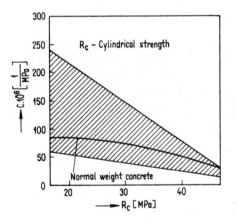

Fig. 1.32 Limits of creep for lightweight concretes

strengths [1.30] while the oblique curve represents the creep of the reference concrete prepared from natural aggregate. According to Fig. 1.32, creep of concretes with higher strengths is larger than, or equal to that of ordinary concrete. However, the effect of different types of artificial aggregate on creep should be investigated separately.

With reference to [1.1], a conservative approach should be adopted when no tests data for the chosen artificial porous aggregate are available, and the following formula is recommended:

$$\varepsilon_{cl} = 1.6\varphi_o\varepsilon_{ec}$$

Here, φ_o is the creep coefficient and ε_{ec} is the elastic strain of normal-weight aggregate concrete having the same cube strength as the lightweight concrete. For

37

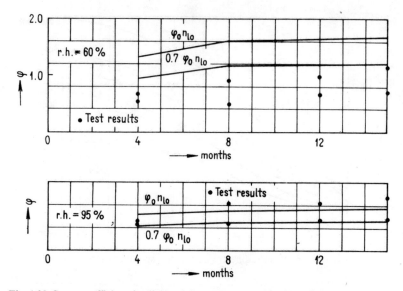

Fig. 1.33 Creep coefficient for lightweight concrete made with agloporite aggregate

example, the modulus of elasticity of a lightweight-aggregate concrete is 0.185×10^5 MPa for a strength of 33 MPa, while that of the normal-weight concrete of the same strength is 0.305×10^5 MPa. Hence, the creep coefficient φ_l of the lightweight-aggregate concrete is approximately $0.97\varphi_o$ and therefore $\varepsilon_{cl} = 0.97\varphi_o\varepsilon_{el}$, where ε_{el} is the elastic strain as given by the modulus of elasticity of the lightweight-aggregate concrete. In [1.32], it is recommended to calculate

$$\varphi_l = \varphi_o \frac{E_{cl}}{E_{co}} = \varphi_o n_{lo}$$

Hence

$$\varepsilon_{cl} = \varphi_l \frac{\sigma}{E_{cl}}$$

where the subscripts o and l designate normal-weight and lightweight concretes, respectively.

Then, for the above case, $\varepsilon_{cl} = 0.61\varphi_o \dfrac{\sigma}{E_{cl}}$. In the literature, however, low values are also proposed, i.e.

$$\varphi_l = 0.7\varphi_o \frac{E_{cl}}{E_{co}}$$

which yields

$$\varepsilon_{cl} = 0.42\varphi_o \frac{\sigma}{E_{cl}}$$

38

Since $\varphi = \dfrac{\varepsilon_c}{\varepsilon_e}$, the foregoing recommendations evidently differ as to the value of the ratio of creep and elastic strain.

Tests on a lightweight-aggregate concrete of strength of 33 MPa, made from agloporite [1.33], yielded low creep coefficients φ; the results are presented in Fig. 1.33 for 60 and 95 per cent relative humidities. A comparison of creep coefficients φ calculated from the measured values, with the lowest values recommended in [1.32], indicates that the relation $\varphi_l = \varphi_o n_{lo}$ is closer to the measured values φ_l for a very high humidity; in drier ambient conditions, the measured coefficients φ were even lower than those resulting from the expression $0.7\varphi_o n_{lo}$.

Measurements of the creep of lightweight concretes show that the effect of different porous artificial aggregates may lead to different values of creep. Hence, it is impossible to formulate any conclusions which would apply for all types of lightweight concrete; therefore, before using concretes made from artificial porous aggregates, the magnitude of creep must be verified by tests.

Water/cement ratio

The water/cement ratio (w/c) has a significant effect on the magnitude of creep, and hence, some Codes take it into account for the analysis of creep. This is necessary especially when concrete is prepared with a very low w/c ratio for a particular type of production, for example, in the precast concrete industry.

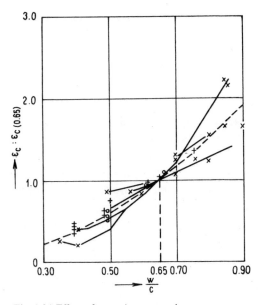

Fig. 1.34 Effect of water/cement ratio on creep

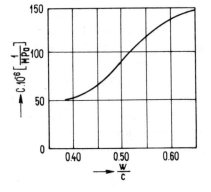

Fig. 1.35 Effect of water/cement ratio on magnitude of final specific creep

The effect of the w/c ratio is evident from Fig. 1.34, which summarizes the research results of several authors [1.8]; creep of concrete with w/c equal to 0.65 is chosen as the reference. The functional dependence of the ratio $\varepsilon_c/\varepsilon_{c(0.65)}$ on w/c is depicted by dashed curve so that the higher the value of w/c, the greater is the creep. Some other test results [1.22], given in Fig. 1.35, are depicted by a smooth curve characterizing the growth of the final value of specific creep with an increase in w/c ratio.

The results presented by the individual authors are difficult to compare: the values in Fig. 1.34 have been modified by a reduction of the measured creep to that of a concrete with the cement paste representing 20 per cent of the mass of concrete. It is apparent that the curves in both figures have similar trends and both relationships agree satisfactorily within the w/c range of $\langle 0.5; 0.65\rangle$; hence, the dependence of creep on the w/c ratio can be determined with a sufficient reliability. The Recommendations, which account for this effect, assume a linear relationship, because more complex relations would encumber the calculation without necessarily making it more accurate.

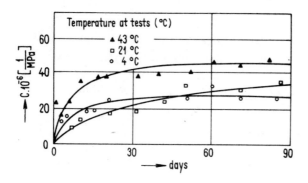

Fig. 1.36 Effect of temperature on the magnitude of specific creep in members subjected to bending

Temperature and curing of concrete

The effect of temperature on creep is frequently neglected in the Standards, although tests have proved that it is not negligible. Under normal conditions, when the temperature varies between $-10\,°C$ and $40\,°C$, the effect of temperature on creep is not pronounced, mainly because the structures are only exposed to extreme temperatures for short periods, and such short-term variations of temperature do not have any significant effect. This effect, however, must be considered for special structures (bunkers containing hot materials, etc.), and especially for heat-cured concretes after accelerated hardening. The effects of temperature on creep then may be divided into those acting on concrete under service conditions, and those influencing concrete in the course of its preparation.

Figure 1.36 [1.34] shows the distribution and magnitude of specific creep during a period of 90 days when concrete members were exposed to various temperatures.

The large decrease of specific creep with the fall of ambient temperature is evident. In these tests, creep was determined by subtracting shrinkage measured on members without load, from the total strain. Figure 1.37 [1.35] indicates the results of creep tests under pressure in which the surface of the specimens was sealed to prevent the passage of moisture between the concrete and the ambiance; the graph shows the zones of specific creep, within which the results varied with varying ambient temperature. It may be observed that the scatter of the results increases with increasing temperature and that, after 100 days, the relationship is almost linear for temperatures exceeding 27 °C; the large effect of temperature on creep is clearly demonstrated.

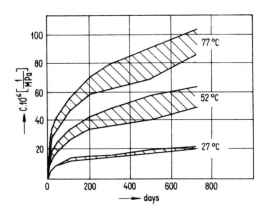

Fig. 1.37 Effect of temperature on the magnitude of specific creep in compressed members

Creep is strongly influenced by steam-curing and autoclaving, both methods accelerating the hardening of concrete. In either case, creep is smaller than that obtained with normal curing of concrete. According to [1.36], creep is reduced by 30 to 50 per cent when concrete is steam-cured for 13 hours at a temperature of 65 °C, the reason being that steam-curing accelerates the hydration, and when subsequently exposed to drier and cooler ambient conditions, dehydration is accelerated. Specific creep and its time-dependent distribution for normally cured concrete and for concrete steam-cured at 66 °C for 13 hours is shown in Fig. 1.38 [1.36].

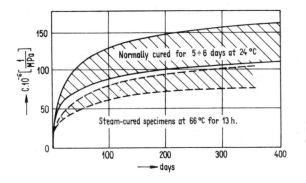

Fig. 1.38 Effect of steam-curing on specific creep; the specimens loaded at a strength of 28.0 MPa

41

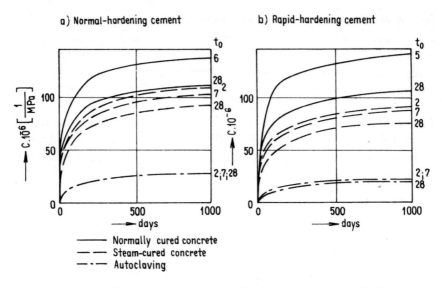

Fig. 1.39 Effect of steam-curing on specific creep (t_0 = age at loading)

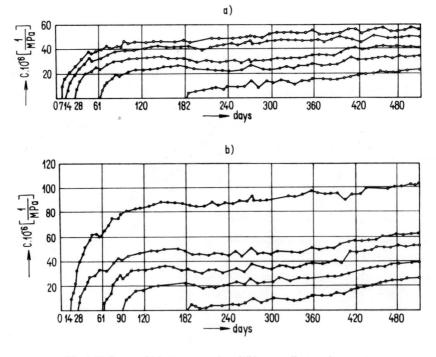

Fig. 1.40 Creep of (a) steam-cured and (b) normally cured concrete

Creep is even further reduced by autoclaving: Fig. 1.39 [1.37] compares creep of normally cured concrete (solid lines), steam-cured concrete (dashed lines) and autoclaved concrete (dot-and-dash lines); Fig. 1.39(a) refers to Portland cement and Fig. 1.39(b) refers to rapid-hardening Portland cement concretes.

Creep tests carried out in the USSR are presented in Fig. 1.40 for steam-cured specimens, and normally cured specimens. It appears that the largest effect of steam-curing occurs in young concrete (up to a 50 per cent reduction), and that this effect decreases with increasing age at loading.

Influence of the state of stress

Tensile stress. Apart from creep of concrete subjected to compressive stresses, other types of stresses have to be considered. However, the number of tests reported is limited as yet, and consequently the information on creep under various types of stress is scarce.

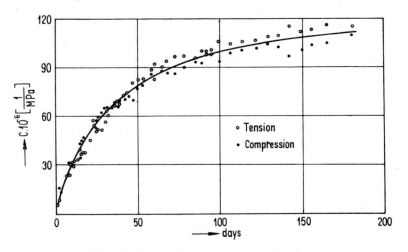

Fig. 1.41 Creep under compression and tension

Most of the collected data refer to creep under concentric tension, although tests for this type of loading are intricate (because of the difficulty of providing uniaxial tension free from eccentricity). According to test data, the development of creep in concrete subjected to tension is basically identical with that in compression, although, perhaps, creep in tension develops quicker during the first months (see Fig. 1.41 [1.38]). In this figure, the circles designate specific creep in tension and the dots designate specific creep in compression; all specimens were subjected to a stress of 1.1 MPa.

Practically no reliable information exists on creep subjected to tension with bending, which is important with regard to the development of cracks. Some tests

43

indicate that the creep characteristic is probably analogous [1.39] to that under compression; however, it should be remembered that the strain gradient is important in this case. The analogy applies also in the case of creep in torsion, which can be considered as a combined effect of tensile and compressive stresses; this is confirmed by reported test results [1.40].

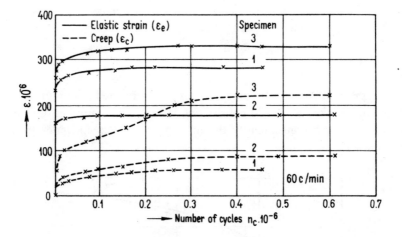

Fig. 1.42 Relation between creep and elastic strain and the number of loading cycles n_c

Repeated load. Permanent strain of concrete increases under the effect of repeated loading, which may also be considered as creep; in this case, as in the case of creep under static loading, its magnitude depends on the age (t_0) at application of load. The reported tests are so few that they do not allow a reliable estimate of the final magnitude of creep under repeated loading. The tests quoted in [1.41] have yielded the following creep coefficients ($\varphi = \varepsilon_c/\varepsilon_e$) after a specified number of loading cycles:

age of concrete (t_0)	8 weeks	10 weeks	7.5 months	31 months
$\varphi(t_0, n_c)$	1.47	1.0	0.2	0.06

The symbol t_0 in the coefficient φ designates the age of concrete at the initial introduction of stress, and the symbol n_c designates the total number of cycles.

Figure 1.42 demonstrates the stabilization of creep after a certain number of cycles [1.42]. The members had different levels of elastic strain ε_e; the stress during the cycles was always compressive with its maximum value equal to 0.47 of the compressive strength; the loading varied at a rate of 60 cycles/min. Evidently, creep developed smoothly under lower stresses (or strains ε_e), whereas, under higher stresses (specimen No. 3), the creep rate suddenly increased after approximately

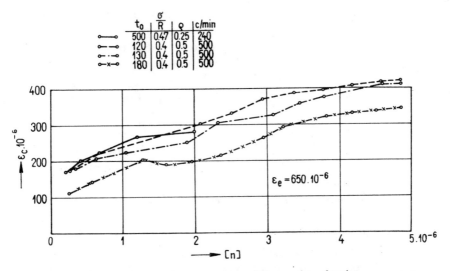

t_0	$\frac{\sigma}{R}$	ϱ	c/min
500	0.47	0.25	240
120	0.4	0.5	500
130	0.4	0.5	500
180	0.4	0.5	500

$\varepsilon_e = 650.10^{-6}$

Fig. 1.43 Relation between creep and the number of cycles

0.05×10^6 cycles, and then subsequently stabilized at approximately 0.3×10^6 cycles. On the other hand, tests reported in [1.43, 1.44] suggest (Fig. 1.43) that creep stabilizes after one million of cycles; the maximum creep was approximately 400×10^{-6} for a value of $\varepsilon_e = 650 \times 10^{-6}$, and the coefficient $\varphi(t_0, n_c)$ was 0.62 for $t_0 = 120$ days; these results agree satisfactorily with those given in [1.41]. Maximum and minimum σ in the tests reported in [1.43, 1.44] were 0.4 and 0.2 of the cube strength of concrete, respectively; n_c was 500 c/min. Similar tests with similar results were carried out in the USSR [1.47]; these results are represented by the solid line in Fig. 1.43.

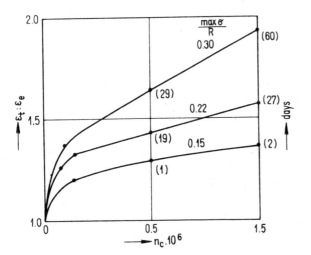

Fig. 1.44 Effect of the number of cycles on the ratio of total strain to elastic strain; numbers in brackets indicate the time at which the ratio $\varepsilon_t/\varepsilon_e$ under a static stress (σ_{max}) is the same as that under repeated loading

45

The magnitude of creep probably depends also on the velocity with which the lowest and the highest stresses alternate. Comparison of the results presented in Figs. 1.42 and 1.43 would suggest that creep stabilizes earlier when the load cycles at a lower rate. The total creep strains from repeated and static loadings are compared in [1.45], and the test results are depicted in Fig. 1.44; here, the ratio of the total and elastic strains at different levels of the ratio of stress–strength ratio (σ/R_c) are plotted on the ordinate. The numbers in parentheses represent the periods of time taken for the total strain from the static load $\sigma = \max \sigma$ to reach the same value of the total strain from the repeated load. Further tests showed that, after 0.72×10^6 cycles, creep from repeated loading was equal to that from static load acting for 28 days; creep after 3.6×10^6 cycles corresponded to 180 days of static loading, and creep after 10.1×10^6 cycles to 600 days of static loading. The tests presented in [1.43] yielded similar results; here, the creep from a static load acting for 100 days corresponded to the average creep after 2×10^6 cycles of repeated load.

It is suggested that the magnitude of creep from repeated loading after 5×10^6 cycles may attain a value up to 0.8 of the elastic strain ε_e; this ratio was 0.66 in the tests presented in Fig. 1.43.

From the small number of tests reported, it is difficult to assess the magnitude of creep from this type of loading. The fundamental cause of the phenomenon is yet more obscure than in the case of static load. It seems, however, that creep under these conditions is caused by the extrusion of water due to the periodic increase and decrease of stress. With this in mind, however, the standing assumption that the final value of total creep from static and repeated loadings is actually equal to the sum of both these individual effects, is still to be clarified.

Modulus of elasticity

The modulus of elasticity of concrete, the elastic strain ε_e and creep ε_c change with the strength of concrete. In fact, the strength of concrete R_c and the modulus of elasticity E_c develop in time like creep. The increase of cube strength R_c is portrayed in Fig. 1.45 [1.18], and its value for a concrete made from normal Portland cement may be determined, for example, from the relation

$$R_{c(t)} \doteq R_{c(28)} \frac{4}{3 + \dfrac{28}{t}} \tag{1.12}$$

or

$$1.36 \frac{t}{t + 10} R_{c(28)}$$

where t is the age of concrete in days.

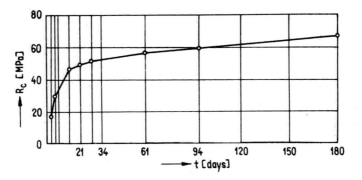

Fig. 1.45 Relation of cubic strength to the age of concrete

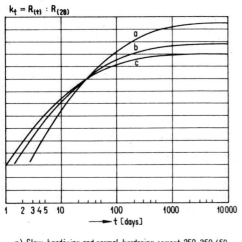

a) Slow-hardening and normal-hardening cement 250, 350, 450
b) Rapid-hardening cement 350, 450
c) Rapid-hardening high-strength cement 550

Fig. 1.46 Relation $k = R_{(t)}/R_{(28)}$ to the age of concrete

Another relationship between the strength and age of concrete is shown in Fig. 1.46 [1.4], in which the type of cement is seen to be influential.

The time-related variation of the modulus of elasticity E_c is shown in Fig. 1.47 [1.18]; here, the modulus of elasticity was determined for a stress corresponding to 0.3 of the strength of $10 \times 10 \times 40$ cm test specimens. Tests made by other authors are shown in Fig. 1.48: the following relationships (t in days) correspond to their individual results:

(a) $E_{c(t)} = 2 \times 10^4 (1 - e^{-0.03t})$ [MPa]
(b) $E_{c(t)} = 3 \times 10^4 (1 - 0.4 e^{-0.06t})$ [MPa]
(c) $E_{c(t)} = 3.73 \times 10^4 (1 - 0.478 e^{-0.017\,5t})$ [MPa]
(d) $E_{c(t)} = 2 \times 10^4 (1 - e^{-0.09t})$ [MPa]

47

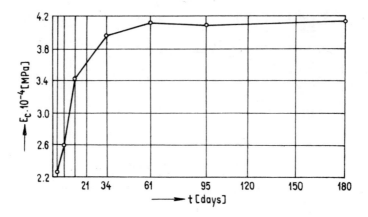

Fig. 1.47 Change of the modulus of elasticity in time

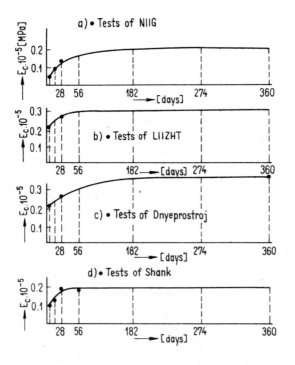

Fig. 1.48 Change of the modulus of elasticity in time according to different tests

Evidently, these relationships can be generalized in the form:

$$E_{c(t)} = E_{c(\infty)} (1 - A_0 \, e^{-B_0 t})$$

where $E_{c(\infty)}$ is the ultimate value of the modulus of elasticity (for $t \to \infty$), and A_0, B_0 are constants depending on the type of concrete. Another formula for $E_{c(t)}$ is [1.5]:

48

$$E_{c(t)} = 14\,000 \left(\frac{R_{c(28)} \times 10}{0.75 + \dfrac{7}{t}} \right)^{\frac{1}{2}} \tag{1.13}$$

This expression is related to the 28-day cube strength $R_{c(28)}$ [MPa], which is more convenient for the calculation than the previous formulae employing the unknown modulus $E_{c(\infty)}$ for $t \to \infty$.

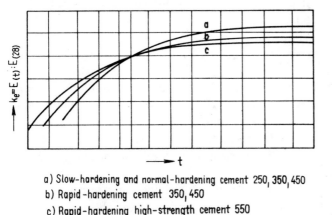

a) Slow-hardening and normal-hardening cement 250, 350, 450
b) Rapid-hardening cement 350, 450
c) Rapid-hardening high-strength cement 550

Fig. 1.49 Effect of age on the modulus of elasticity when concrete is cured under normal conditions

Alternatively, the value of modulus of elasticity may be determined at any time with the aid of the relationship shown in Fig. 1.49 [1.4], where the type of cement is also considered.

If the modulus of elasticity E_c is known at the age of 28 days, the following relationship may also be used to determine $E_{c(t)}$

$$E_{c(t)} = \left(1.36 \frac{t}{t + 10} \right)^{\frac{1}{2}} E_{c(28)}$$

1.2.2 Shrinkage and swelling of concrete

The reason for these volume changes of concrete is that the cement gel either shrinks or increases its volume. This occurs when chemically free water evaporates from concrete in a dry environment (shrinkage), or when concrete absorbs water from a humid environment (swelling). For these processes, it is immaterial whether concrete is loaded or not, but the capillary pressure is important. Shrinkage and swelling depend on the ambient conditions, the age of concrete, the mix composition and the method of preparation, the dimensions of the structural member, and on other factors similar to those considered for creep. Both shrinkage and swelling are functions of time. Many of these factors have been introduced into the Standards and Recommendations; the Soviet Codes, for example, express shrinkage by means of coefficients in the formula

$$\varepsilon_s = \varepsilon_{s\infty} \eta_1 \eta_2 \eta_4$$

where η_1 and η_2 have the same meaning as for creep, $\varepsilon_{s\infty}$ is the final value of shrinkage under various climatic conditions, and η_4 is a coefficient accounting for the age of concrete (t_0). Similar coefficients are also introduced in the CEB-FIP Recommendations (see Sect. 2.6.1).

Effect of time and of the age of concrete

At constant temperature and humidity, shrinkage under dry conditions develops quickly at first, and subsequently decreases until it stabilizes at its ultimate value $\varepsilon_{s\infty}$. In water, where concrete increases its volume, the growth of strain is similar.

The development of shrinkage under constant hygrometric conditions is shown in Fig. 1.50 [1.21], while Fig. 1.51 [1.51] portrays this development under different ambient conditions.

Under laboratory conditions with controlled humidity and temperature, the shrinkage curves are on the whole smooth. However, in practice, both the relative humidity and temperature vary so that the rate of development of shrinkage changes accordingly. This is a consequence of a partial reversibility of shrinkage: if a specimen is moved from a dry ambient condition to a humid condition, not only is shrinkage arrested, but swelling may develop if the specimen is stored in

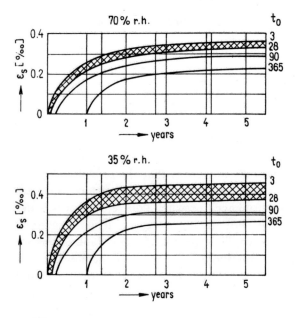

Fig. 1.50 Effect of the medium; t_0 is the time at which the specimens were stored under humid conditions and from which shrinkage is calculated

water. This process is depicted in Fig. 1.51 which shows the test results for specimens stored in free air (70% relative humidity, 18 °C), then in water, and then in free air again.

Structures subjected to actual ambient conditions undergo this process; owing to the variations of humidity (and temperature), shrinkage turns into a phenomenon which never terminates. The results of shrinkage data started at different ages of some observed structural members are shown in Fig. 1.52 [1.43]. The measurements (dashed lines) commenced 5 months ($\tau = 150$ days) after casting for members A, after 60 days for members B and after one day for members C.

The other two curves were calculated in accordance with the Mörsch–Dischinger relationship, and the variation of humidity and temperature is depicted in the bottom part of the diagram.

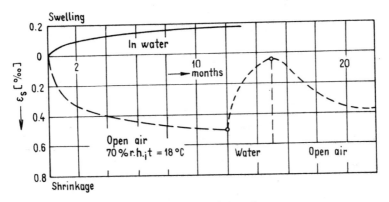

Fig. 1.51 Shrinkage and swelling of concrete

The relationship in Fig. 1.51 also demonstrates the time-dependent distribution and magnitude of shrinkage in relation to the age τ of concrete: the final value of shrinkage measured on older concrete is lower than on a younger concrete. The tests shown in Fig. 1.51 also prove the following: if the specimens remain under humid conditions until the commencement of measurements (i.e. they are adequately cured), and then stored under dry conditions, shrinkage develops rapidly at first but slows down later. This development corresponds with that observed, when concrete is stored in free air after a temporary immersion in water (Fig. 1.51): concrete shrinks rapidly in the first period and the temporary immersion in water causes a decrease of the magnitude of ultimate shrinkage.

Investigations into the effect of age on the magnitude of shrinkage, as carried out by several authors, are summarized in Fig. 1.53 [1.23]; here, the effect of the age (i.e. time t) is expressed by the coefficient η_4 (see Sect. 2.6.1). Although the test results are scattered, curve A is seen to represent the generalized effect of age.

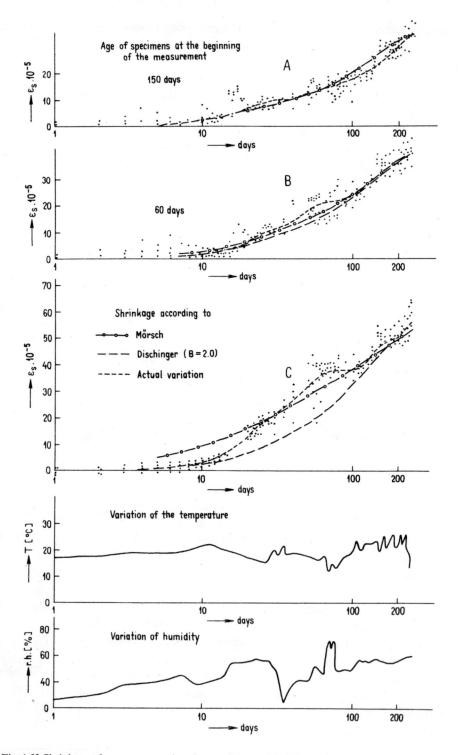

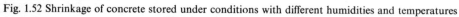

Fig. 1.52 Shrinkage of concrete stored under conditions with different humidities and temperatures

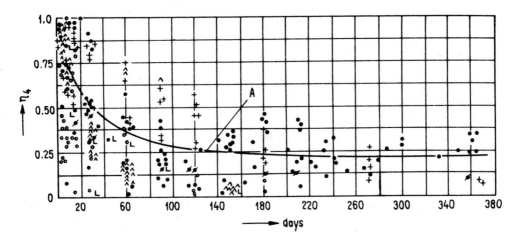

Fig. 1.53 Effect of the age on shrinkage

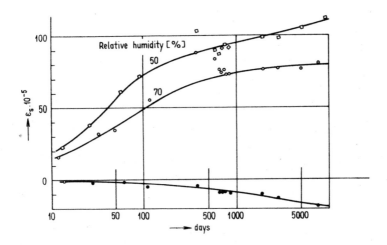

Fig. 1.54 Effect of the ambient conditions on shrinkage and swelling

Ambient conditions

The influence of constant ambient conditions on the development and magnitude of shrinkage is depicted in Fig. 1.50, where the effects of 70 and 35 per cent relative humidities may be compared. Results for 50 and 70 per cent relative humidities, and for specimens stored in water, are depicted in Fig. 1.54 [1.4], where it appears that the rate of shrinkage depends on relative humidity; shrinkage of a specimen stored in 50 per cent relative humidity did not stabilize in the test period.

Tests performed by some authors and summarized in Fig. 1.55 [1.43] demonstrate the relationship between final values of shrinkage and relative humidity: for ⟨80; 40⟩

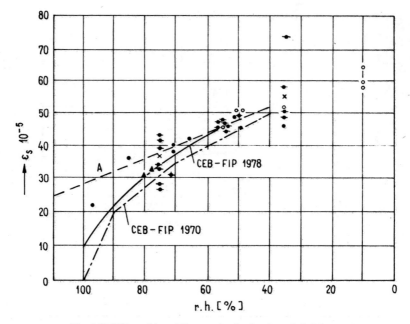

Fig. 1.55 Effect of humidity on the final value of shrinkage

per cent relative humidity (r.h.), the following linear relationship (A) is acceptable:

$$\varepsilon_{s\infty} = 15.0 + 0.37\,(110 - \text{r.h.} \%) \times 10^{-5}$$

The figure also gives the curve resulting from the CEB-FIP Recommendations.

Dimensions of the structural member

It has been demonstrated that the hydrometric state of the ambient conditions exerts a large influence on shrinkage, In addition, its magnitude is influenced by the surface of the member exposed to the environment.

A two-fold approach is adopted here, as in the case of creep: either the effect of the size is related to the minimum dimension of the member ($d_{\min}$ [1.23]), or to the effective thickness $d_f = 2A/O$ (Sect. 2.6); the latter method is adopted in further discussions as it is considered to be more accurate.

The effect of the dimensions on the magnitude of shrinkage is portrayed in Fig. 1.56 [1.43] which contains test results of several authors; the coefficient k'_e (refer to Sect. 2.6.2) is the ordinate and the minimum dimension of the member ($d_{\min}$) is the abscissa. Curve A represents the average of the measured values [1.23], while the other two curves specify the zone in accordance with the Recommendations [1.1]; the top curve applies for a square section ($a = b$) and the bottom curve is for a plate-shaped section ($b \to \infty$).

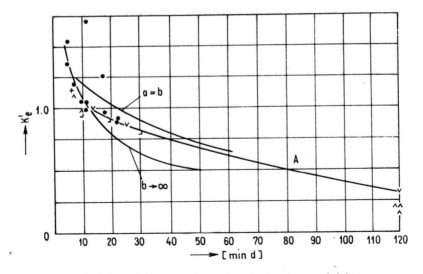

Fig. 1.56 Effect of the dimensions (d_f) of the member on shrinkage

Water/cement ratio, type of cement and temperature

The water/cement ratio exerts a large influence on shrinkage: the higher the ratio, the greater is shrinkage. The reason for this is that the strength of concrete, and especially its porosity depend to a large extent on the w/c ratio. The higher the w/c ratio, the more porous is the concrete and the more favourable are the conditions for a more intensive interchange of moisture between the concrete and the ambiance. The relationship between shrinkage and w/c ratio is depicted in Fig. 1.57 [1.19], the curves of which were obtained from test data.

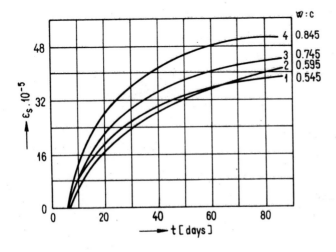

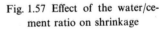

Fig. 1.57 Effect of the water/cement ratio on shrinkage

55

In practice, the influence of *w/c* ratio on shrinkage is taken into account in codes and standards by using the same values as for creep.

The type of cement has a small effect on shrinkage (the fineness of grinding is the most important factor), hence it is not considered in the analyses.

The effect of temperature is similar to that for creep.

2
Theoretical and practical prediction
of creep and shrinkage

2.1 Formulae for the prediction of creep

As noted earlier, the creep strains of concrete due to a constant stress σ acting from time τ until t, may be assumed to be proportional to the elastic strains, according to the relation (compare also with Eq. (1.10))

$$\varepsilon_c(t, \tau) = \varphi(t, \tau)\,\varepsilon_e = C(t, \tau)\,\sigma \tag{2.1}$$

where the function $\varphi(t, \tau)$, expressing the ratio of the creep strain to the instantaneous elastic strain under a constant stress, is called the creep coefficient. The function $C(t, \tau)$ is designated as the specific creep, which represents the creep strain at age t caused by a unit sustained stress acting since age τ ($\tau < t$). Characterization of creep by $\varphi(t, \tau)$ is frequently a source of confusion since the strains are thus unnecessarily made dependent upon $E(\tau)$, which is usually taken not as the truly instantaneous (dynamic) modulus, but as the modulus corresponding to the strain in a short time interval after load application (for further details, see [2.1]).

The stress produced strain (mechanical strain) is often expressed as

$$\varepsilon_\sigma(t,\tau) = \varepsilon_e + \varepsilon_c = \sigma J(t, \tau)$$

in which $J(t, \tau)$ is the compliance function (the creep function) representing the strain at time t caused by a unit constant stress that has been acting since time τ. According to Eq. (2.1) it can be written

$$J(t, \tau) = \frac{\varepsilon_e + \varepsilon_c}{\sigma} = \frac{1}{E(\tau)} + C(t, \tau) = \frac{1 + \varphi(t, \tau)}{E(\tau)}$$

where $E(\tau)$ is the elastic modulus characterizing the instantaneous strain at age τ.

The analytical expression of the functions φ and C agrees approximately with the results of tests. The recommended relationships should fit the available experimental data for the types of concretes considered and take into account all the important factors (age, temperature, environmental humidity, size and shape of cross-section, and curing conditions). The unknown coefficients of the functions should be relatively easy to evaluate from the available experimental or empirical data [2.2]. At the same time, the functions should be sufficiently simple to facilitate the numerical evaluation of practical calculations.

In the historical development of research into creep, specific creep $C(t, \tau)$ is usually explained on the basis of one of three fundamental theories, commonly known as the theory of delayed elasticity, the rate-of-creep theory, and the general theory.

2.1.1   The theory of delayed elasticity

It is assumed that the age τ of concrete does not affect the final value of the specific creep, and hence, creep depends only on the duration of the loading, i.e. on the time period $(t-\tau)$. After unloading, creep for $t \to \infty$ vanishes completely. Specific creep is determined by the expression

$$C(t - \tau) = C(\infty)\left[1 - e^{-B_1(t - \tau)}\right] \tag{2.2}$$

Hence in this theory, concrete is considered a material whose creep is not affected by age. In Eq. (2.2), $C(\infty)$ is the value of specific creep for $t \to \infty$ and B_1 is a constant; both of them depend on the properties of concrete.

The shape of the specific creep $C(t - \tau)$ curve, shown in Fig. 2.1a, is the same for any age at loading; the curves for individual ages are displaced in the t-direction, and have the same value for $t \to \infty$. When the load is removed at time τ_2

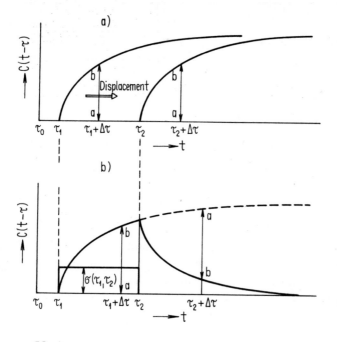

Fig. 2.1 Creep according to the delayed elasticity theory

58

(Fig. 2.1b), a reverse process occurs, again in agreement with Eq. (2.2). Evidently, this function of creep is actually the delayed elastic strain as discussed in Section 1.1; Eq. (1.6) also corresponds basically to Eq. (2.2). The rheological model is that of Kelvin for a viscoelastic material (Fig. 1.4), and when $E = E(t)$, the model for total creep (ε) is that shown in Fig. 2.4a.

If $C(\infty) = \dfrac{1}{E}\,\varphi(\infty)$, then in accordance with Eq. (2.2)

$$C(t - \tau) = \frac{\varphi(\infty)}{E}\left[1 - e^{-B_1(t - \tau)}\right] = \frac{\varphi}{E} \tag{2.3}$$

where $\varphi = \varphi(\infty)\left[1 - e^{-B_1(t - \tau)}\right]$.

If concrete is under constant stress for the entire period of time, and if t approaches infinity ($t \rightarrow \infty$), then the strain is

$$\varepsilon(\infty) = \frac{\sigma(\tau_1)}{E} + \sigma(\tau_1)\,C(\infty) =$$

$$= \frac{\sigma(\tau_1)}{E}\left[1 + \varphi(\infty)\right] = \frac{\sigma(\tau_1)}{E_r} \tag{2.4}$$

where

$$E_r = \frac{E}{1 + \varphi(\infty)} \tag{2.5}$$

Consequently, the final strain can be calculated with the aid of the effective modulus E_r (see also Sect. 2.4.3.6).

A comparison of test results with the creep calculated by means of the delayed elasticity theory indicates that this theory is not suitable for concretes loaded at an early age; it is considered unsuitable for concretes loaded earlier than one year, because, during this period, the age of concrete has a considerable effect on the final creep value and on the irreversible creep, both of which are not taken into account in this theory.

For concretes loaded at a later age, agreement of results with the theory of viscoelastic material is more satisfactory. For this reason, it is sometimes recommended to apply the delayed elasticity method to calculate creep in old concretes.

2.1.2 The rate-of-creep theory

In this theory, the effect of the age of concrete on the final value of creep is taken into account; specific creep is given by the relationship

$$C(t, \tau) = C(t) - C(\tau) \tag{2.6}$$

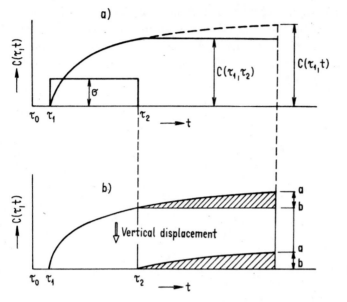

Fig. 2.2 Creep according to the rate-of-creep theory

and with a constant modulus of elasticity E, it becomes

$$C(t, \tau) = \frac{1}{E} \left[\varphi(t) - \varphi(\tau) \right] \qquad (2.7)$$

Specific creep at time t then depends on the age at the instant at which stress was applied to the concrete; all curves may be obtained from the original one (for $\tau = 0$) by its displacement in the vertical direction. If the structural member is unloaded at time τ_2, then, in agreement with the principle of superposition, the portion of the curve related to time τ_2 is subtracted from the original curve, so as to maintain the strain which exists in the concrete at time τ_2 (Fig. 2.2a). Evidently, this theory neglects the reversible part of the strain, and no delayed creep recovery is predicted. Moreover, for old concrete negligible creep is obtained, which is incorrect according to experimental results.

Here, concrete has the character of a viscous material with non-linear viscosity as depicted in the model shown in Fig. 2.4b.

The creep coefficient $\varphi(t)$ is expressed by means of various formulae to achieve maximum agreement between the analytical results and test data; the following expressions are commonly used:

$$\varphi(t) = \varphi(\infty)(1 - e^{-B_2 t}) \qquad \text{(Dischinger)} \qquad (2.8a)$$

$$\varphi(t) = \varphi(\infty)(1 - e^{-t^{\frac{1}{2}}})^{\frac{1}{2}} \qquad \text{(Mörsch)} \qquad (2.8b)$$

$$\varphi(t) = \varphi(\infty)\frac{t}{B_2' + t} \qquad \text{(Ross)} \qquad (2.8c)$$

60

The final values of the creep coefficient $\varphi(\infty)$ and the constants B_2, B'_2 are determined from tests.

The difference between the concept of concrete as a viscoelastic and ageing material may be demonstrated by comparing Eqs. (2.3) and (2.8a); in the first case

$$C(t - \tau) = \frac{\varphi(\infty)}{E} [1 - e^{-B_1(t - \tau)}]$$

while in accordance with the rate-of-creep theory

$$C(t, \tau) = \frac{\varphi(\infty)}{E} (e^{-B_2\tau} - e^{-B_2 t}) =$$

$$= \frac{\varphi(\infty)}{E} e^{-B_2\tau} [1 - e^{-B_2(t - \tau)}]$$

here, $\dfrac{\varphi(\infty)}{E} e^{-B_2\tau}$ is the coefficient of affinity, i.e. the ratio of the values $C(t, \tau)$ with the same interval $(t - \tau)$, but with different ages of concrete τ.

The expressions in the brackets are identical, except for the constants; for a viscoelastic material, the multiplier is only the constant $\dfrac{\varphi(\infty)}{E}$, while, for an ageing material, the multiplier also depends on the age at loading.

If t approaches infinity, according to the rate-of-creep theory for a constant load which was applied at time τ_1:

$$\varepsilon(\infty) = \frac{\sigma(\tau_1)}{E} + \sigma(\tau_1) C(t, \tau_1) = \frac{\sigma(\tau_1)}{E} [1 + \varphi(\infty) e^{-B_2\tau_1}] =$$

$$= \frac{\sigma(\tau_1)}{E_r}$$

where

$$E_r = \frac{E}{1 + \varphi(\infty) e^{-B_2\tau_1}}$$

Hence, when $\tau_1 \to \infty$, then $E_r \to E$, and $\varphi(\infty) e^{-B_2\tau_1} \to 0$ so that the concrete behaves elastically. Test data, however, prove also that concretes loaded at a greater age exhibit creep deformations, and consequently, the reality lies between the two theories discussed so far.

2.1.3 The general theory

The mean compliance function of the cross-section as a whole in the presence of drying may be expressed approximately as

$$J(t, \tau) = J_b(t, \tau) + C_d(t, \tau) \tag{2.9a}$$

where $J_b(t, \tau)$ is the compliance function for constant pore humidity and thermal state and $C_d(t, \tau)$ is the mean additional compliance due to drying.

The component $J_b(t, \tau)$ can be written in the form

$$J_b(t, \tau) = \frac{1}{E_0} + C_b(t, \tau) \qquad (2.9b)$$

in which the first term represents the elastic deformation. The second term represents the creep of concrete at constant humidity and temperature (the basic creep), which may be well described by power curves of load duration $(t - \tau)$, and by inverse power curves for the effect of age τ at loading. This is called the double power law

$$C_b(t, \tau) = \frac{\Phi_1}{E_0} (\tau^{-m} + \alpha) (t - \tau)^n \qquad (2.9c)$$

in which $n \simeq 1/8$, $m \simeq 1/3$, $\alpha \simeq 0.05$, $\Phi_1 \simeq 3$ to 6 (if τ and t are in days), and E_0 ($=$ asymptotic modulus) $\simeq 1.5$ times the conventional elastic modulus for concrete aged 28 days [2.2].

It has been found that the double power law exhibits certain deviations from experimental data. An improvement can be obtained by the recently proposed triple power law [2.21] — the corresponding compliance function is given by the expression

$$J_b(t, \tau) = \frac{1}{E_0} + \frac{\Phi_1}{E_0} (\tau^{-m} + \alpha) (t - \tau)^n - B(t, \tau, n)$$

in which

$$B(t, \tau, n) = n \int_{\xi=0}^{t-\tau} \left[1 - \left(\frac{\tau}{\tau + \xi} \right)^n \right] \xi^{n-1} \mathrm{d}\xi$$

$$\xi = t - \tau$$

Function $B(t, \tau, n)$, representing the deviation from the double power law, is a binomical integral. A table of values of this integral has been presented in [2.21].

The total creep strain has been separated into the basic and drying components (Eq. (2.9a)). The basic creep term involves no effect of humidity and no effect of the size of the cross-section. These effects are involved only in the drying creep term $C_d(t, \tau)$.

The drying creep term in the BP model for prediction of creep [2.19] (see Sect. 2.6.5) is given basically as a function of humidity, the duration of loading and the shrinkage-square half time, which represents the dependence of drying creep on the size of the cross-section. This also brings in the effect of diffusivity, along with its dependence on the age of concrete at the start of drying.

Note that the models utilizing the environmental humidity rather than the local pore humidities are incapable of providing stress and strain distributions throughout the cross-sections. Their purpose is the use in calculations of the distributions of bending moments and deflections along beams in structures (see [2.22]).

Many other expressions for the mean specific creep have been proposed. The models made no distinction between the basic creep and the drying creep. These models include the expression

$$C(t, \tau) = C(\infty, \tau) \left[1 - e^{-B_3(t - \tau)} \right] \tag{2.10}$$

where $C(\infty, \tau)$ is the final value of specific creep, which decreases with an increase in age τ. Hence, $C(\infty, \tau)$ expresses the creep magnitude at $t \to \infty$.

Usually, the term $C(\infty, \tau)$ is expressed by the relationship

$$C(\infty, \tau) = A_1 + \frac{A_2}{\tau} \tag{2.11a}$$

or

$$C(\infty, \tau) = A_1 + A_2 e^{-B_3\tau} \tag{2.11b}$$

where A_1, A_2 and B_3 are constants determined from tests.

The structure of $C(\infty, \tau)$ in Eq. (2.10) indicates that it is a combination of both the theories presented in Sections 2.1.2 and 2.1.3, because, for example, using the relationship (2.11b)

$$C(t, \tau) = A_1 \left[1 - e^{-B_3(t - \tau)} \right] + A_2 e^{-B_3\tau} \left[1 - e^{B_3(t - \tau)} \right] \tag{2.10a}$$

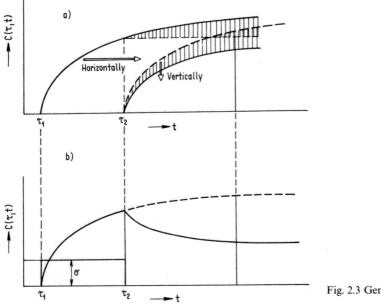

Fig. 2.3 General theory

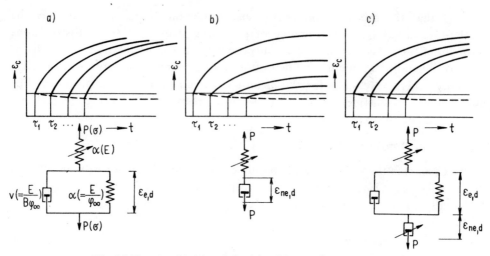

Fig. 2.4 Theories: (a) delayed elasticity; (b) rate-of-creep; (c) general

The resulting expression is then obtained by a horizontal displacement of the curve in agreement with the delayed elasticity theory, and by its vertical displacement in agreement with the rate-of-creep theory (Fig. 2.3a). After unloading (Fig. 2.3b), part of the strain is reversible, but part of the strain is of a permanent character. The model expressing the total strain is depicted in Fig. 2.4c.

To establish the best possible agreement with test data, specific creep may be expressed in a more general form by the Dirichlet series [2.2], [2.3], [2.6], [2.28], [2.29].

$$C(t, \tau) = \sum_{i=1}^{m} C_i(\infty, \tau) \left[1 - e^{-B_{3,i}(t-\tau)} \right] \tag{2.12}$$

Equation (2.10) corresponds to a single term ($i = 1$) of this series.

Equation (2.12) may be employed to express all distributions of specific creep with any required accuracy. The coefficients C_i and $B_{3,i}$ for the required m terms of the series are best found by means of, for example, the method of least squares, usually with the the aid of a computer [2.2].

2.1.4   Comparison of theories

Each of the introduced theories has its characteristic time-shape for the total creep which can be represented by a rheological model (shown in Fig. 2.4). The latter theory expressed by Eq. (2.10a) leads to the rate-of-creep (with $A_1 = 0$) and delayed elasticity (with $A_2 = 0$) theories.

The general theory can approximate the actual evaluation of the final creep value as well as its time distribution. If the principle of superposition applies, creep under a variable stress, or changes of stress arising from a prescribed strain can be theoretically calculated. When choosing one of the theories for practical calculations, the burden of the solution should be considered, and also the necessity of a time-consuming analysis with regard to the inaccuracies involved in the choice of the input data (such as the time-variable environmental conditions, composition of the mix, etc.); in addition, there is the variability of the constants which appear in the formulae. Consequently, it is frequently more reasonable to choose a "less exact" method which yields simpler solutions so that the calculation of creep is facilitated.

It has been demonstrated that the theory of delayed elasticity is unsuitable for the solutions of creep of concretes loaded soon after casting; it is better suited for older concretes, i.e. for those loaded not before the age of 1.5 years, when nearly all the strains are reversible; of course, a realistic specific creep $C(\infty)$ would have to be chosen.

On the other hand, the rate-of-creep method might be acceptable for calculating creep of concretes loaded earlier than one year after hardening, this being the most frequent case in practice. However, the deficiency of this method is underestimated. Relaxation is also less (zero at $\tau \to \infty$) and is thus underestimated in the calculations. Hence, in relaxation-type problems, prediction by this method represents an upper limit of the stress change, [2.2].

2.1.5 Modified rate-of-creep method

The rate-of-creep theory has up to now been widely used. Since it has many shortcomings, especially in completely neglecting the recoverable strain, methods are sought which utilize its relatively simple solutions in relaxation-type problems, while being more accurate. A modification of the rate-of-creep theory has been introduced [2.4] which divides creep into inelastic, irrecoverable, creep $\varepsilon_{ne,d}$ (viscous strain) and delayed elastic strain $\varepsilon_{e,d}$ corresponding to a viscoelastic material (see Chap. 1). The method is a hybrid of the delayed elasticity and the rate-of-creep methods. Its predictions usually lie between these two methods and are thus closer to the exact solution.

The total strain, in accordance with Eq. (1.7), without the effect of shrinkage, is (Fig. 2.5):

$$\varepsilon_t = \varepsilon_e + \varepsilon_{e,d} + \varepsilon_{ne,d} \tag{2.14}$$

It is assumed that the age of concrete at which there is a change of stress affects the magnitude and rate of the delayed elastic strain ($\varepsilon_{e,d}$) through the influence of the modulus of elasticity $E(\tau)$. Then it may be written:

$$\varepsilon_{e,d} = \sigma(\tau)\, C_{e,d}(t - \tau) = \sigma(\tau)\frac{\varphi_{e,d}(t - \tau)}{E(\tau)} = \sigma(\tau)\frac{\varphi_1(t - \tau)}{E(\tau)} \tag{2.15a}$$

65

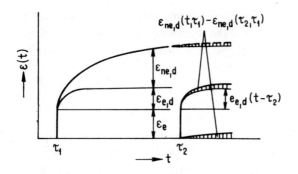

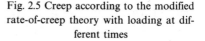

Fig. 2.5 Creep according to the modified rate-of-creep theory with loading at different times

The rate-of-creep theory applies to the irrecoverable part of the delayed strain ($\varepsilon_{ne,d}$), and consequently, the following relationship holds:

$$\varepsilon_{ne,d} = \sigma(\tau)\left[C_{ne,d}(t) - C_{ne,d}(\tau)\right] =$$
$$= \sigma(\tau)\frac{\varphi_{ne,d}(t) - \varphi_{ne,d}(\tau)}{E(\tau)} = \sigma(\tau)\frac{\varphi_2(t) - \varphi_2(\tau)}{E(\tau)} \tag{2.15b}$$

To simplify Eq. (2.15) the following designation has been introduced:

$$\varphi_{e,d} = \varphi_1; \qquad \varphi_{ne,d} = \varphi_2$$

The above expression satisfies the principle of superposition, and the functions $\varphi_{e,d}$ and $\varphi_{ne,d}$ may be determined so that the predicted results correspond to reality; the solution may be considered to be as accurate as the solution obtained by the general theory so that the rheological model of this theory is analogous to that shown in Fig. 2.4c. The only difference is that the strains $\varepsilon_{e,d}$ and $\varepsilon_{ne,d}$ are not mutually separated in the general theory.

The effect of the age of concrete on the magnitude of the total strain is portrayed in Fig. 2.5; if a stress $\sigma(\tau_1)$ has been applied at instant τ_1, the strain at time t is

$$\varepsilon(t) = \varepsilon_e + \varepsilon_{e,d}(t - \tau_1) + \varepsilon_{ne,d}(t, \tau_1)$$

if a stress $\sigma(\tau_2)$ is applied at a subsequent instant τ_2, according to the assumptions of the rate-of creep theory

$$\varepsilon(t) = \varepsilon_e + \varepsilon_{e,d}(t - \tau_2) + \varepsilon_{ne,d}(t, \tau_1) - \varepsilon_{ne,d}(\tau_2, \tau_1)$$

Evidently, in this case, the elastic delayed strain does not depend on τ_1.

When the member is unloaded at time τ_2, the strain at instant $t > \tau_2$ is (Fig. 2.6)

$$\varepsilon(t) = \varepsilon_e + \varepsilon_{e,d}(t - \tau_1) + \varepsilon_{ne,d}(t, \tau_1) - \varepsilon_e -$$
$$- \varepsilon_{e,d}(t - \tau_2) - \varepsilon_{ne,d}(t, \tau_1) + \varepsilon_{ne,d}(\tau_2, \tau_1) =$$
$$= \varepsilon_{e,d}(t - \tau_1) - \varepsilon_{e,d}(t - \tau_2) + \varepsilon_{ne,d}(\tau_2, \tau_1)$$

66

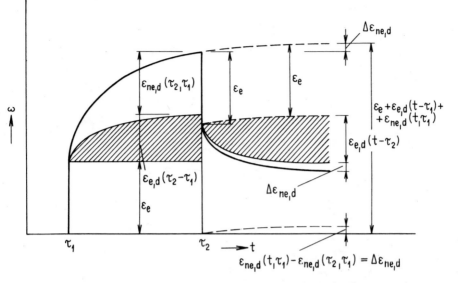

Fig. 2.6 Modified rate-of-creep theory; creep after unloading

The first two terms $\varepsilon_{e,d}$ correspond to the theory of delayed elasticity and the last term to the rate-of-creep theory.

In accordance with Eq. (2.15), the values of the creep strains $\varepsilon_{e,d}$ and $\varepsilon_{ne,d}$ at instant t depend (as in the general theory) on the final values of $\varepsilon_{e,d\infty}$ and $\varepsilon_{ne,d\infty}$ and, considering $E(\tau) = \text{constant} = E$, it may be written

$$\varepsilon_{e,d} = \sigma(\tau)\frac{\varphi_{1\infty}k_1(t - \tau)}{E} \tag{2.16a}$$

$$\varepsilon_{ne,d} = \sigma(\tau)\frac{\varphi_{2\infty}[k_2(t) - k_2(\tau)]}{E} \tag{2.16b}$$

The limiting values of $\varphi_{1\infty}$ and $\varphi_{2\infty}$, as well as the time-distributions of $k_1(t - \tau)$ and $k_2(t)$, have to be determined.

The main problem is then the magnitude and distribution of $\varepsilon_{e,d}$ which have to be determined by tests. It is shown from these tests that the delayed elastic strain develops more rapidly after stress application than the strain $\varepsilon_{ne,d}$, while $\varepsilon_{ne,d}$ attains its final values after several years; on the other hand, the majority of the strain $\varepsilon_{e,d}$ terminates within approximately 100 days. An example illustrating the development of both these strains is shown in Fig. 2.7; to obtain the lines A to C, concrete specimens of the same age are loaded and gradually unloaded at times τ_1, $\tau_2, \ldots, \tau_i$ (Fig. 2.7a) so that the measured strains are taken from the base lines aa', bb', cc', thus determining the development of the elastic delayed strain. If the final values of $\varepsilon_{e,d}$ and $\varepsilon_{ne,d}$ at their respective times are considered, then the

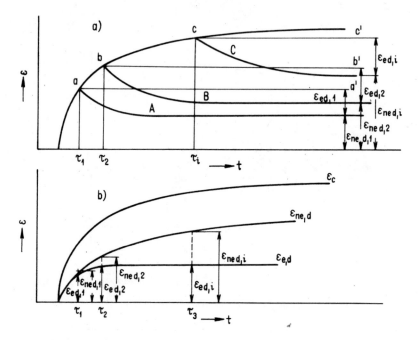

Fig. 2.7 Delayed elastic strain $\varepsilon_{e,d}$, irreversible strain $\varepsilon_{ne,d}$ and total strain ε_t

relationship of Fig. 2.7b is obtained for every instant. It may be seen that the final value of $\varepsilon_{e,d}$, for the same stress, is determined by the duration of the loading $(\tau_i - \tau_0)$. Identical tests, but with varying age (τ_0), indicate that the age of concrete at loading has only little effect on the strain $\varepsilon_{e,d}$; also, for the same concrete, its final value is almost independent of its age (τ_0) at loading. The dependence of the magnitude of the ratio $\varepsilon_{e,d}/\varepsilon_{e,d\infty} = k_1$ on the duration of loading $(t - \tau)$ is represented in Fig. 2.8. [2.4] in which the test results demonstrate that 70 per cent of this value was reached approximately after 100 days.

However, even the final value of the ratio $\varepsilon_{e,d\infty}/\varepsilon_e$ is not subjected to any large variations. Some results are shown below; tests were carried out with the ratio $\sigma/R_{bk} = 0.35$ [2.4] on concretes of varying cement content and w/c ratio:

content of cement [kg/m³]	1 010	590	435	322
w/c	0.31	0.44	0.49	0.53
$\varepsilon_{e,d\infty}/\varepsilon_{e28}$	0.25	0.28	0.32	0.41

There is still not enough information available for a reliable determination of the elastic delayed strain: for the time being, it is proposed to consider $\varphi_{1\infty} = \varepsilon_{e,d\infty}/\varepsilon_{e28} = 0.4$ in the analyses. This value is somewhat higher than that cor-

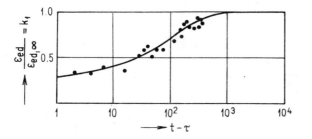

Fig. 2.8 Ratio of the elastic delayed
strain to its final value $\varepsilon_{ed.\,r}$

responding to the results for concretes of typical composition and stress/strength ratios, but it contains an allowance which is necessary in the analysis of creep. Using this value, in accordance with Eq. (2.14), then with σ = constant, in Eqs. (2.16a) and (2.16b)

$$\varepsilon(t) = \frac{1}{E}\,\sigma(\tau_1)\,\{1 + 0.4k_1(t - \tau_1) + \varphi_{2\infty}[k_2(t) - k_2(\tau_1)]\} \qquad (2.17)$$

This expression yields higher values $\varepsilon(t)$ than if $\varphi_{1\infty}$ were less than 0.4.

It is recommended that the function $k_1(t - \tau)$ be obtained from test data, for example, those of Fig. 2.8 where the limiting value approaches unity.

The final value of the coefficient $\varphi_{2\infty}$ must be determined so that the sum of strains $\varepsilon_{e,d\infty}$ and $\varepsilon_{ne,d\infty}$ corresponds to the test results; the same applies to the distribution of k_2. Some recommended values of all the coefficients are given in Sect. 2.6.

The modified rate-of-creep theory does not lead to a more simplified analysis than the general theory. In the evaluation of strains from a known stress history, the effects are merely summed in both theories, but analysis of relaxation from a known strain history involves the same disadvantages for both theories. For this reason, it is proposed to treat the delayed elastic component as constant, $\varepsilon_{e,d} = 0.4$ which corresponds to a simplification in accordance with Fig. 2.8. Equation (2.17) then reads

$$\varepsilon(t) = \frac{1}{E}\,\sigma(\tau_1)\,\{1.4 + \varphi_{2\infty}[k_2(t) - k_2(\tau_1)]\} \qquad (2.18)$$

Obviously, the solution is approximate but the advantage of this theory is that it allows a crude analysis with the assumption of a constant delayed elastic strain, or a more accurate solution, when its actual time-development is considered.

2.1.6 Creep under variable stress

An assumption of linearity implies validity of the principle of superposition. This principle states that the strain response due to the sum of stress histories is the sum of the individual responses. Hence, if concrete is subjected to a uniaxial

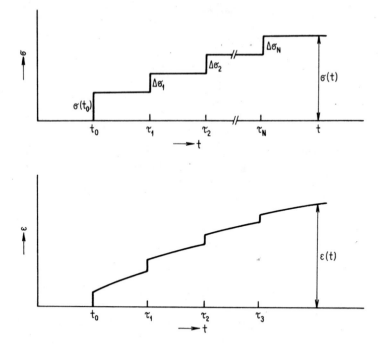

Fig. 2.9 Stress history and the corresponding strain variation

stress which changes its magnitude abruptly at times τ_1, τ_2, etc. (see Figs. 2.9 and 1.13), then the strain of concrete in time t is described by the relation

$$\varepsilon(t) = \frac{\sigma(t_0)}{E(t_0)} + \sigma(t_0)\, C(t, t_0) + \sum_{i=1}^{N} \Delta\sigma_i(\tau_i)\left[\frac{1}{E(\tau_i)} + C(t, \tau_i)\right] \qquad (2.19)$$

If the creep coefficient is used in the expression instead of the specific creep C, Eq. (2.19) becomes

$$\varepsilon(t) = \frac{\sigma(t_0)}{E(t_0)}\left[1 + \varphi(t, t_0)\right] + \sum_{i=1}^{N} \Delta\sigma_i(\tau_i)\frac{1 + \varphi(t, \tau_i)}{E(\tau_i)} \qquad (2.20)$$

If the stress history $\sigma(\tau)$ is continuous, summing the strain histories due to all small stress increments before time t, changes Eq. (2.19) into

$$\varepsilon(t) = \frac{\sigma(t_0)}{E(t_0)} + \sigma(t_0)\, C(t, t_0) + \int_{t_0}^{t} \frac{d\sigma(\tau)}{d\tau}\left[\frac{1}{E(\tau)} + C(t, \tau)\right] d\tau \qquad (2.21)$$

and Eq. (2.20) into

$$\varepsilon(t) = \frac{\sigma(t_0)}{E(t_0)}\left[1 + \varphi(t, t_0)\right] + \int_{t_0}^{t} \frac{d\sigma(\tau)}{d\tau}\left[\frac{1}{E(\tau)} + \frac{\varphi(t, \tau)}{E(\tau)}\right] d\tau \qquad (2.22)$$

In both these relations, the strain $\varepsilon(t)$ is expressed by means of the initial stress $\sigma(t_0)$. Sometimes, however, it is more convenient to relate the strain to the stress which acts in the structural member at instant t; this can be achieved by integrating Eq. (2.21) by parts. Then, we have

$$\varepsilon(t) = \frac{\sigma(t_0)}{E(t_0)} + \sigma(t_0)\,C(t, t_0) + \int_{t_0}^{t} \frac{d\sigma(\tau)}{d\tau}\left[\frac{1}{E(\tau)} + C(t, \tau)\right] =$$
$$= \frac{\sigma(t_0)}{E(t_0)} + \sigma(t_0)\,C(t, t_0) + \left\{\sigma(\tau)\left[\frac{1}{E(\tau)} + C(t, \tau)\right]\right\}_{t_0}^{t} -$$
$$- \int_{t_0}^{t} \sigma(\tau)\frac{\partial}{\partial\tau}\left[\frac{1}{E(\tau)} + C(t, \tau)\right]d\tau \tag{2.23}$$

Since $C(t, t) = 0$, this relation may be written in the form of

$$\varepsilon(t) = \frac{\sigma(t_0)}{E(t_0)} + \sigma(t_0)\,C(t, t_0) + \frac{\sigma(t)}{E(t)} - \sigma(t_0)\left[\frac{1}{E(t_0)} + C(t, t_0)\right] -$$
$$- \int_{t_0}^{t} \sigma(\tau)\frac{\partial}{\partial\tau}\left[\frac{1}{E(\tau)} + C(t, \tau)\right]d\tau \tag{2.24}$$

which leads to the relation

$$\varepsilon(t) = \frac{\sigma(t)}{E(t)} - \int_{t_0}^{t} \sigma(\tau)\frac{\partial}{\partial\tau}\left[\frac{1}{E(\tau)} + C(t, \tau)\right]d\tau \tag{2.25}$$

The first part of this relation represents the elastic strain related to time t, and the second part expresses the development of strain during the observed time interval $\langle t_0, t \rangle$; the latter strain depends on the stress in concrete at each instant, i.e. on the loading history.

For practical use, the relations are often, sometimes approximately, simplified by introducing the assumption of a constant modulus of elasticity (i.e. $E(\tau) = $ = constant = E). Then (2.22) is written

$$\varepsilon(t) = \frac{1}{E}\left\{\sigma(t_0)\left[1 + \varphi(t, t_0)\right] + \int_{t_0}^{t} \frac{d\sigma(\tau)}{d\tau}\left[1 + \varphi(t, \tau)\right]d\tau\right\} \tag{2.26}$$

and with the introduction of the creep coefficient, Eq. (2.25) becomes

$$\varepsilon(t) = \frac{1}{E}\left\{\sigma(t) - \int_{t_0}^{t} \sigma(\tau)\frac{\partial}{\partial\tau}\left[1 + \varphi(t, \tau)\right]d\tau\right\} \tag{2.27}$$

The fundamental equations (2.21) and (2.25) are accurate within the limits of the introduced assumptions; the accuracy of the analysis, i.e. the agreement of predicted and actual strains, depends on the choice of the functional relationships used to express specific creep $C(t, \tau)$ (or the creep coefficient $\varphi(t, \tau)$ and the modulus of elasticity $E(\tau)$).

2.2   Theoretical expression for shrinkage

Shrinkage, which develops from instant τ until instant t, is usually expressed by the relation

$$\varepsilon_s(t) = \varepsilon_{s\infty}[k_s(t) - k_s(\tau)] \tag{2.28}$$

where $\varepsilon_{s\infty}$ is the final value of shrinkage which depends on the ambient conditions, the size of the member and the w/c ratio; k_s expresses the shrinkage function; t designates the time to which shrinkage is considered, and τ is the age of concrete from which shrinkage is observed. The coefficient k_s is usually governed by an exponential relationship of the form

$$k_s = (1 - e^{-Bt^C})^D \tag{2.29}$$

Usually, $C = D = 1$, and if $B = B_2$, then

$$k_s = 1 - e^{-B_2 t} \quad \text{(see Eq. (2.8a))}$$

Mörsch's diagram of the time-distribution of shrinkage is based on the assumption that $B = 1$ and $C = D = 0.5$; then

$$k_s = (1 - e^{-t^{0.5}})^{0.5} \quad \text{(see Eq. (2.8b))}$$

2.3   Practical methods of calculating strain from a known stress history

When the loading history is known, the method to be used in the calculation of the strain of concrete at time t depends especially on the adopted expression of the creep coefficient $\varphi(t, \tau)$. The following methods are particularly applicable:

2.3.1   Direct integration

This method is based on the evaluation of values $\varepsilon(t)$ by a direct application of Eqs. (2.21), (2.22), (2.25) through (2.27) by carrying out indicated integrations. However, this is possible only in special cases of stress histories and with the use of special forms of the expression for the creep coefficient. From the practical point of view, the rate-of-creep method gives the most convenient analytical solutions (Sect. 2.1.2), for example, using Eq. (2.7) with a constant modulus of elasticity, Eq. (2.27) may be written in the simple form

$$\varepsilon(t) = \frac{1}{E}\left[\sigma(t) + \int_{t_0}^{t} \sigma(\tau)\frac{d\varphi(\tau)}{d\tau}\,d\tau\right] \tag{2.30}$$

Using the modified rate-of-creep theory according to Sect. 2.1.5, Eq. (2.22), with consideration of Eqs. (2.15a) and (2.15b), acquires the form

$$\varepsilon(t) = \frac{\sigma(t_0)}{E(t_0)} + \frac{\sigma(t_0)}{E(t_0)} \varphi_1(t - t_0) + \frac{\sigma(t_0)}{E(t_0)} [\varphi_2(t) - \varphi_2(t_0)] +$$

$$+ \int_{t_0}^{t} \frac{d\sigma(\tau)}{d\tau} \left[\frac{1}{E(\tau)} + \frac{\varphi_1(t - \tau)}{E(\tau)} + \frac{\varphi_2(t) - \varphi_2(\tau)}{E(\tau)} \right] d\tau \qquad (2.31)$$

With a constant modulus of elasticity, Eqs. (2.26) and (2.27) read

$$\varepsilon(t) = \frac{1}{E} \sigma(t_0) [1 + \varphi_1(t - t_0) + \varphi_2(t) - \varphi_2(t_0)] +$$

$$+ \frac{1}{E} \int_{t_0}^{t} \frac{d\sigma(\tau)}{d\tau} [1 + \varphi_1(t - \tau) + \varphi_2(t) - \varphi_2(\tau)] d\tau \qquad (2.32)$$

$$\varepsilon(t) = \frac{\sigma(t)}{E} - \frac{1}{E} \int_{t_0}^{t} \sigma(\tau) \frac{\partial}{\partial \tau} [1 + \varphi_1(t - \tau) + \varphi_2(t) - \varphi_2(\tau)] d\tau \qquad (2.33)$$

The last relation may be further simplified when a constant value of the delayed elastic strain (equal to 40 per cent of the elastic strain — Eq. (2.18)) is introduced:

$$\varepsilon(t) = \frac{\sigma(t)}{E} - \frac{1}{E} \int_{t_0}^{t} \sigma(\tau) \frac{\partial}{\partial \tau} [1.4 + \varphi_2(t) - \varphi_2(\tau)] d\tau \qquad (2.34)$$

It is, however, evident from the structure of this equation that by differentiating with respect to the variable τ, the constants 1.4 and $\varphi_2(t)$ disappear and the relation becomes identical with Eq. (2.30). Thus, the influence of delayed elasticity ceases to exist.

2.3.2 Method of time-discretization — summation according to Eq. (2.20)

The realistic forms of creep law usually do not permit analytical solutions, and so numerical techniques must be employed.

When the strain at time t is calculated according to the time-discretization method, it is assumed that the time period t_0, t is subdivided into a number of time intervals within which the stress can be considered constant (Fig. 2.9). This is the simplest method which allows the use of any expression for the creep coefficient $\varphi(t, \tau)$, and hence, it is very convenient for an analysis involving programable pocket calculators.

73

Writing Eq. (2.20) successively for a growing i, we obtain

$$
\begin{Bmatrix} \varepsilon_0 \\ \varepsilon_1 \\ \varepsilon_2 \\ \vdots \\ \varepsilon_N \end{Bmatrix} =
\begin{bmatrix}
\dfrac{1}{E_0}, & 0, & 0, & & \\
D_{1,0}, & \dfrac{1}{E_1}, & 0, & & \\
D_{2,0}, & D_{2,1}, & \dfrac{1}{E_2}, & & \\
\vdots & \vdots & \vdots & \vdots & \\
D_{N,0}, & D_{N,1}, & D_{N,2} & \cdots, & D_{N,N-1}, & \dfrac{1}{E_N}
\end{bmatrix}
\begin{Bmatrix} \sigma_0 \\ \Delta\sigma_1 \\ \Delta\sigma_2 \\ \vdots \\ \Delta\sigma_N \end{Bmatrix}
\tag{2.35}
$$

where

$$E_i = E(\tau_i)$$

$$D_{k,i} = \frac{1 + \varphi(t_k, \tau_i)}{E(\tau_i)} \qquad\qquad \sigma_0 = \sigma(t_0) \tag{2.36}$$

Equations (2.35) can be written in matrix form

$$\{\varepsilon\} = [D]\{\Delta\sigma\}$$

The advantage of this and the following methods is that they do not have any special demands on the expression for the stress history which may even be discontinuous, this being a typical case in concrete structures (such as a change of the structural system, a sudden change of loading of the structure, subsidence of a support, etc.).

2.3.3 Method of time-discretization — summation according to Eq. (2.25)

The time period $\langle t_0, t \rangle$ is subdivided into k time intervals of a duration Δt $\left(\Delta t = \dfrac{t - t_0}{k} \right)$.

Equation (2.25), using the creep coefficient $\varphi(t, \tau)$, is

$$\varepsilon(t) = \frac{\sigma(t)}{E(t)} - \int_{t_0}^{t} \sigma(\tau) \frac{\partial}{\partial\tau} \left\{ \frac{1}{E(\tau)} [1 + \varphi(t, \tau)] \right\} d\tau \tag{2.38}$$

when summation is substituted for the integral (see also [2.5])

$$\varepsilon(t) = \frac{\sigma(t)}{E(t)} - \sum_{j=0}^{k-1} \int_{t_0 + j\Delta t}^{t_0 + (j+1)\Delta t} \sigma(\tau) \frac{\partial}{\partial\tau} \left\{ \frac{1}{E(\tau)} [1 + \varphi(t, \tau)] \right\} d\tau \tag{2.39}$$

If the stress within the individual time interval is assumed constant, i.e.

$$\sigma(\tau) = \sigma(t_0 + j\Delta t) = \sigma_j \tag{2.40}$$

Eq. (2.39) may be written in the form

$$\varepsilon(t) = \frac{\sigma(t)}{E(t)} + \sum_{j=0}^{k-1} \sigma_j F_j \tag{2.41}$$

where

$$F_j = - \int_{t_0+j\Delta t}^{t_0+(j+1)\Delta t} \frac{\partial}{\partial \tau} \left\{ \frac{1}{E(\tau)} [1 + \varphi(t, \tau)] \right\} d\tau =$$

$$= - \left\{ \frac{1}{E(\tau)} [1+\varphi(t, \tau)] \right\}_{t_0+j\Delta t}^{t_0+(j+1)\Delta t} = \tag{2.42}$$

$$= \frac{1}{E(t_0 + j\,\Delta t)} - \frac{1}{E[t_0 + (j+1)\,\Delta t]} +$$

$$+ \frac{\varphi(t, t_0 + j\,\Delta t)}{E(t_0 + j\,\Delta t)} - \frac{\varphi[t, t_0 + (j+1)\,\Delta t]}{E[t_0 + (j+1)\,\Delta t]}$$

This formula may be expressed more conveniently as

$$F_j = \frac{1}{E_j} - \frac{1}{E_{j+1}} + \frac{\varphi_j}{E_j} - \frac{\varphi_{j+1}}{E_{j+1}} \tag{2.43}$$

In the case of a constant modulus of elasticity, the expression is more simplified:

$$F_j = \frac{1}{E} [\varphi(t, t_0 + j\,\Delta t) - \varphi(t, t_0 + (j+1)\,\Delta t)]*)$$

Writing Eq. (2.41) successively for a growing j, we obtain

$$
\left\{
\begin{array}{c}
\varepsilon_0 \\
\varepsilon_1 \\
\varepsilon_2 \\
\vdots \\
\varepsilon_k
\end{array}
\right\} =
\left[
\begin{array}{ccccc}
\frac{1}{E_0}, & 0, & 0, & & \\
F_0, & \frac{1}{E_1}, & 0, & & \\
F_0, & F_1, & \frac{1}{E_2}, & & \\
\vdots & \vdots & \vdots & \vdots & \\
F_0, & F_1, & F_2, F_3, & \ldots, F_{k-1}, & \frac{1}{E_k}
\end{array}
\right]
\left\{
\begin{array}{c}
\sigma_0 \\
\sigma_1 \\
\sigma_2 \\
\vdots \\
\sigma_k
\end{array}
\right\} \tag{2.44}
$$

which may be written in matrix form

$$\{\varepsilon\} = [F] \{\sigma\}$$

This method again is very convenient for programable pocket calculators.

---

*) Applying the rate-of-creep theory, it becomes

$$F_j = \frac{1}{E} [\varphi(t_0 + (j+1)\,\Delta t) - \varphi(t_0 + j\Delta t)]$$

Both of these foregoing methods represent a numerical solution of the same problem and are mutually equivalent. If the investigated time period is divided (for the sake of simplicity), for example, only in two time intervals, it follows from Eq. (2.20), or from Eq. (2.35) (where $N = 2$), that

$$\varepsilon(t) = \sigma(t_0)\frac{1 + \varphi(t, t_0)}{E(t_0)} + \Delta\sigma_1\frac{1 + \varphi(t, \tau_1)}{E(\tau_1)} + \Delta\sigma_2\frac{1}{E(t)} \tag{2.46}$$

and, with reference to Eq. (2.44), we obtain

$$\varepsilon(t) = \sigma(t_0)\left[\frac{1}{E(t_0)} - \frac{1}{E(\tau_1)} + \frac{\varphi(t, t_0)}{E(t_0)} - \frac{\varphi(t, \tau_1)}{E(\tau_1)}\right] +$$

$$+ \sigma(\tau_1)\left[\frac{1}{E(\tau_1)} - \frac{1}{E(t)} + \frac{\varphi(t, \tau_1)}{E(\tau_1)}\right] + \sigma(t)\frac{1}{E(t)} =$$

$$= \frac{\sigma(t_0)}{E(t_0)}[1 + \varphi(t, t_0)] + \frac{\sigma(\tau_1) - \sigma(t_0)}{E(\tau_1)}[1 + \varphi(t, \tau_1)] +$$

$$+ \frac{\sigma(t) - \sigma(\tau_1)}{E(t)} \tag{2.47}$$

Since the following apply (Fig. 2.9)

$$\sigma(\tau_1) - \sigma(t_0) = \Delta\sigma_1$$
$$\sigma(t) - \sigma(\tau_1) = \Delta\sigma_2$$

Eqs. (2.46) and (2.47) are identical.

2.3.4 Solution using a special form of the creep coefficient

If the creep coefficient is expressed by the relation

$$\varphi(t, \tau) = \sum_{i=1}^{m} \bar{C}_i(\tau)[1 - e^{-\bar{B}_i(t - \tau)}] \tag{2.48}$$

(see Eq. (2.12) in Sect. 2.1.3, and [2.6]), for the stress history depicted in Fig. 2.10, the strain at time t_k, according to Eq. (2.20), is

$$E\varepsilon(t_k) = \sigma(t_0)[1 + \varphi(t, t_0)] + \sum_{j=1}^{k} \Delta\sigma_j[1 + \varphi(t, t_j)] \tag{2.49}$$

This relation may be written, using the expression according to Eq. (2.48), for example, for $k = 4$,

$$E\varepsilon(t_4) = \sigma(t_0)\left\{1 + \sum_{i=1}^{m} \bar{C}_i(t_0)[1 - e^{-\bar{B}_i(\Delta t_4 + t_3 - t_0)}]\right\} +$$

$$+ \Delta\sigma_1\left\{1 + \sum_{i=1}^{m} \bar{C}_i(t_1)[1 - e^{-\bar{B}_i(\Delta t_4 + t_3 - t_1)}]\right\} + \tag{2.50}$$

76

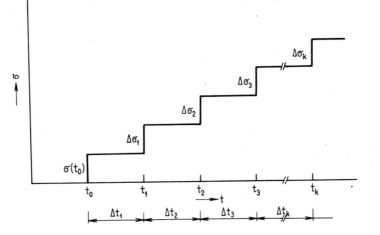

Fig. 2.10 Step-wise variation of stress

$$+ \Delta\sigma_2\{1 + \sum_{i=1}^{m} \bar{C}_i(t_2) [1 - e^{-\bar{B}_i(\Delta t_4 + t_3 - t_2)}]\} +$$

$$+ \Delta\sigma_3\{1 + \sum_{i=1}^{m} \bar{C}_i(t_3) [1 - e^{-\bar{B}_i \Delta t_4}]\}$$

Designating

$$e^{-\bar{B}_i \Delta t_4} = a_{i4} \tag{2.51}$$

we can write

$$E\varepsilon(t_4) = \sigma(t_0)\{1 + \sum_{i=1}^{m} \bar{C}_i(t_0) [1 - a_{i4} e^{-\bar{B}_i(t_3 - t_0)}]\} +$$

$$+ \Delta\sigma_1\{1 + \sum_{i=1}^{m} \bar{C}_i(t_1) [1 - a_{i4} e^{-\bar{B}_i(t_3 - t_1)}]\} +$$

$$+ \Delta\sigma_2\{1 + \sum_{i=1}^{m} \bar{C}_i(t_2) [1 - a_{i4} e^{-\bar{B}_i(t_3 - t_2)}]\} + \tag{2.52}$$

$$+ \Delta\sigma_3\{1 + \sum_{i=1}^{m} \bar{C}_i(t_3) [1 - a_{i4}]\}$$

Since, for $k = 3$

$$E\varepsilon(t_3) = \sigma(t_0)\{1 + \sum_{i=1}^{m} \bar{C}_i(t_0) [1 - e^{-\bar{B}_i(t_3 - t_0)}]\} +$$

$$+ \Delta\sigma_1\{1 + \sum_{i=1}^{m} \bar{C}_i(t_1) [1 - e^{-\bar{B}_i(t_3 - t_1)}]\} + \tag{2.53}$$

$$+ \Delta\sigma_2\{1 + \sum_{i=1}^{m} \bar{C}_i(t_2) [1 - e^{-\bar{B}_i(t_3 - t_2)}]\}$$

77

the strain at time t_4 can be expressed, with the aid of the strain at time t_3, as follows:

$$E\varepsilon(t_4) = E\varepsilon(t_3) + \sigma(t_0) \sum_{i=1}^{m} \bar{C}_i(t_0) e^{-\bar{B}_i(t_3 - t_0)} (1 - a_{i4}) +$$

$$+ \Delta\sigma_1 \sum_{i=1}^{m} \bar{C}_i(t_1) e^{-\bar{B}_i(t_3 - t_1)} (1 - a_{i4}) +$$

$$+ \Delta\sigma_2 \sum_{i=1}^{m} \bar{C}_i(t_2) e^{-\bar{B}_i(t_3 - t_2)} (1 - a_{i4}) +$$

$$+ \Delta\sigma_3 + \Delta\sigma_3 \sum_{i=1}^{m} \bar{C}_i(t_3) (1 - a_{i4})$$

(2.54)

which can be written in the form

$$E\varepsilon(t_4) = E\varepsilon(t_3) + E\Delta\varepsilon_{3,4}$$

(2.55)

where

$$E \Delta\varepsilon_{3,4} = \Delta\sigma_3 + \sum_{i=1}^{m} A_{i4}(1 - a_{i4})$$

(2.56)

and

$$A_{i4} = A_{i3}a_{i3} + \Delta\sigma_3 \bar{C}_i(t_3)$$
$$\vdots$$
$$A_{i1} = \sigma_0(t_0) \bar{C}_i(t_0)$$

(2.57)

The advantage of this procedure is that, in the analysis, the whole loading history need not be considered repeatedly for each successive time interval. To calculate the strain at the i-th time interval it suffices (see Eq. (2.55)) to use the earlier ascertained value in the preceding interval and to correct it by the increment depending on the magnitude of the stress acting in the present time interval and on the coefficient A, which is also easily found from its value for the preceding time interval (see the chain Eq. (2.57)).

2.4 Analysis of stress variation with known strain history, and relaxation

The problem of determining the stress acting on concrete when the strain history is known, is an inverted problem of the calculation of strain at a known stress history, as has been demonstrated in the foregoing Sect. 2.3. If, however, this analysis of the strain is easy (i.e., the integration or the numerical methods demand a minimum mathematical treatment), the determination of the stress at a given strain history is much more difficult. It involves the solution of a Volterra integral

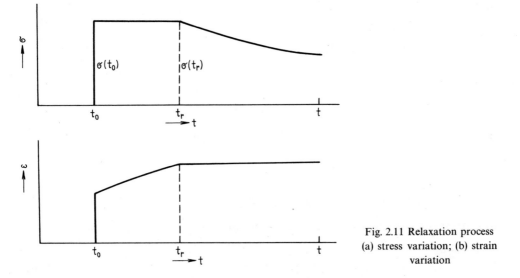

Fig. 2.11 Relaxation process (a) stress variation; (b) strain variation

equation of the type (Eq. (2.22)) with the given function $\varepsilon(t)$ and the required function $\sigma(\tau)$.

The commonest case of this problem is the analysis of relaxation of stress (also refer to Sect. 1.1.2), i.e the investigation of stress variation at constant strain, $\varepsilon(t) = $ constant (Fig. 2.11). This is a typical engineering problem, frequently encountered in the structural analysis.

The numerical analysis of relaxation is only easy when a special expression is used to describe creep.

2.4.1 Analysis according to the theory of delayed elasticity

The stress variations $\sigma(\tau)$ at a known strain can be calculated from the equation

$$\frac{1}{B_1 \varphi(\infty)} \frac{d^2\sigma(t)}{dt^2} + \frac{d\sigma(t)}{dt} \left[\frac{1}{\varphi(\infty)} + 1 \right] =$$

$$= \frac{E}{B_1 \varphi(\infty)} \frac{d^2\varepsilon(t)}{dt^2} + \frac{E}{\varphi(\infty)} \frac{d\varepsilon(t)}{dt} \tag{2.58}$$

The quantities are designated in accordance with Eq. (2.3). The equation may be derived from the model depicted in Fig. 2.4, where the stress is given by

$$\sigma = \varepsilon_1 E = \varepsilon_2 \frac{E}{\varphi(\infty)} + \frac{d\varepsilon_2}{dt} \frac{E}{B_1 \varphi(\infty)}$$

$$\frac{d\varepsilon}{dt} = \frac{d\varepsilon_1}{dt} + \frac{d\varepsilon_2}{dt} \tag{2.59}$$

79

and also

$$\frac{d^2\varepsilon}{dt^2} = \frac{d^2\varepsilon_1}{dt^2} + \frac{d^2\varepsilon_2}{dt^2} \tag{2.60}$$

Here, ε_1 designates the strain on the offset helical spring and ε_2 the strain of the Kelvin model. Integrating Eq. (2.58) we obtain

$$\frac{1}{B_1\varphi(\infty)} \frac{d\sigma(t)}{dt} + \sigma(t)\left[\frac{1}{\varphi(\infty)} + 1\right] = \frac{E}{B_1\varphi(\infty)} \frac{d\varepsilon(t)}{dt} +$$

$$+ \frac{E}{\varphi(\infty)} \varepsilon(t)$$

When relaxation develops from instant t_0 (Fig. 2.12) at which concrete was subjected to a stress $\sigma(t_r)$, the relations $\varepsilon(t) = $ constant and $\dfrac{d\varepsilon(t)}{dt} = 0$ apply, as well as

$$\frac{d\sigma(t)}{dt} + \sigma(t) B_1 [1 + \varphi(\infty)] = EB_1\varepsilon_0 \tag{2.61}$$

The particular solution of Eq. (2.61) can be found by considering $\sigma(t) = \sigma(t_r) = = C_1\varepsilon(t_r)$; then

$$\sigma(t_r) = \frac{E}{1 + \varphi(\infty)} \varepsilon(t_r)$$

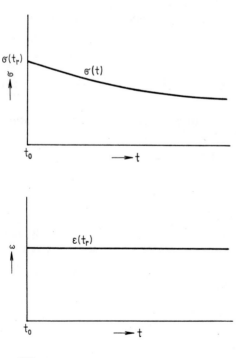

Fig. 2.12 Relaxation process for $t_0 = t_r$

The solution of the homogeneous Eq. (2.61) yields

$$\sigma_1(t) = C_1 \, e^{-B_1[1 + \varphi(\infty)]t}$$

and the final solution is

$$\sigma(t) = C_1 \, e^{-B_1[1 + \varphi(\infty)]t} + \frac{E}{1 + \varphi(\infty)} \, \varepsilon(t_r) \tag{2.62}$$

As the stress $\sigma(t_r)$ acts in the structural member at time $t = 0$ and the corresponding strain is $\varepsilon(t_r) = \dfrac{\sigma(t_r)}{E}$, then

$$\sigma(t_r) = C_1 + \frac{\sigma(t_r)}{1 + \varphi(\infty)}$$

whence

$$C_1 = \sigma(t_r)\left[1 - \frac{1}{1 + \varphi(\infty)}\right]$$

The stress at an arbitrary instant is

$$\sigma(t) = \sigma(t_r)\left\{\frac{1}{1 + \varphi(\infty)} + \left[1 - \frac{1}{1 + \varphi(\infty)}\right] e^{-B_1[1 + \varphi(\infty)]t}\right\} \tag{2.63}$$

The time t is measured from the application of loading which is also the origin of the relaxation process.

2.4.2 Analysis based on the rate-of-creep theory

Differentiating Eq. (2.30) with respect to t, we obtain the differential equation

$$\frac{1}{E}\left[\frac{d\sigma(t)}{dt} + \sigma(t)\frac{d\varphi(t)}{dt}\right] = \frac{d\varepsilon(t)}{dt} \tag{2.64}$$

In the case of relaxation, i.e. with $\varepsilon = $ constant, then

$$\frac{d\varepsilon(t)}{dt} = 0$$

Thus, we have a differential equation

$$\frac{d\sigma(t)}{dt} + \sigma(t)\frac{d\varphi(t)}{dt} = 0 \tag{2.65}$$

whose solution is

$$\sigma(t) = \sigma(t_r)e^{-[\varphi(t) - \varphi(t_r)]} \tag{2.66}$$

where t_r is the time at the beginning of relaxation (Fig. 2.11).

As stated earlier, the rate-of-creep theory does not take into account the effect of previous loading. In the case of the analysis of relaxation problems, this means that, using the rate-of-creep theory, the stress history preceding the commencement of the relaxation process is not influential and hence, neither is the effect of the instant t_0 at the application of loading (Fig. 2.11); this is evident from Eq. (2.66), where only the time t_r of the beginning of relaxation is involved.

It should be noted that with the application of the modified rate-of-creep theory (Sect. 2.1.5) in its simplest form, where the constant delayed elastic strain is introduced by a fixed value equal to 40 per cent of the elastic strain (Eq. (2.18)), Eq. (2.34) for the calculation of relaxation merely yields a differential equation given by Eq. (2.65). The effects of delayed elasticity again are ignored.

If we use the modified rate-of-creep theory in its more generalized form in accordance with Eq. (2.33), where the delayed elastic strain is characterized by the component $\varphi_1(t - \tau)$ of the creep coefficient, and depends on the duration of the time interval $(t - \tau)$, it is possible to analyse the relaxation by applying the procedure with reference to [2.7].

Equation (2.33) may be arranged to the form

$$\varepsilon(t) = \frac{\sigma(t)}{E} - \frac{1}{E} \int_{t_0}^{t} \sigma(\tau) \left[\frac{d\varphi_1(z)}{dz} \frac{dz}{d\tau} - \frac{d\varphi_2(\tau)}{d\tau} \right] d\tau \tag{2.67}$$

where

$$z = t - \tau \tag{2.68}$$

and hence

$$\frac{dz}{d\tau} = -1$$

The stress σ which is variable in the process of relaxation, may be considered to be the sum of two components

$$\sigma(\tau) = \sigma_0 - \sigma_2(\tau) \tag{2.69}$$

where the component σ_0 is the initial stress acting from instant t_0 when concrete was subjected to load. The component σ_0 is constant in time and the variability of the stress is represented by the component $\sigma_2(\tau)$ which starts with zero value at the commencement of the relaxation process at time t_r (Fig. 2.13); consequently, the following initial condition applies:

$$\sigma_2(t_r) = 0 \tag{2.70}$$

Equation (2.67) then may be written in the form

$$\varepsilon(t) = \frac{1}{E} \left\{ \sigma_0 - \sigma_2(t) + \int_{t_0}^{t} [\sigma_0 - \sigma_2(\tau)] \left[\frac{d\varphi_1(z)}{dz} + \frac{d\varphi_2(\tau)}{d\tau} \right] d\tau \right\} \tag{2.71}$$

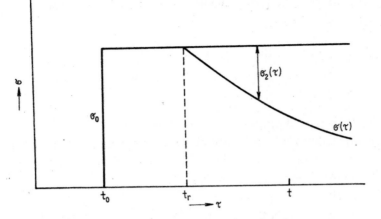

Fig. 2.13 Stress components during relaxation

To avoid the difficulties resulting from the function φ_1 (i.e. not depending directly on time τ but depending on the duration of loading $z = t - \tau$), a linear relationship may be assumed as a first approximation of its shape, i.e.

$$\varphi_1(z) = a + kz \tag{2.72}$$

The gradient k of the time-dependent diagram of the delayed elastic strain may be determined approximately by assuming a dependence on the durations of stress and relaxation:

(a) the stress σ_0 acts from instant t_0 (Fig. 2.13); a part of the delayed elastic strain, caused by this stress and related to time $t_r \leqq \tau \leqq t$ of the duration of the relaxation process, may be characterized by the gradient

$$k_0 = \frac{\varphi_1(t - t_0) - \varphi_1(t_r - t_0)}{t - t_r} \tag{2.73}$$

(b) on the other hand, the stress σ_2 does not appear before the relaxation process. The corresponding gradient, therefore, is to be expressed in a different way, for example,

$$k_2 = \frac{\varphi_1(t - t_r) - \varphi_1(0)}{t - t_r} \tag{2.74}$$

In the course of relaxation, a constant value of strain is maintained. In accordance with Eqs. (2.71) to (2.74), we obtain a differential equation for the distribution of the stress component σ_2 in the form

$$\frac{d\sigma_2(t)}{dt} + \sigma_2(t)\left[k_2 + \frac{d\varphi_2(t)}{dt}\right] = \sigma_0\left[k_0 + \frac{d\varphi_2(t)}{dt}\right] \tag{2.75}$$

83

The general solution of the homogeneous equation of Eq. (2.75) is

$$\sigma_2(t) = Ce^{-[k_2 t + \varphi_2(t)]} \tag{2.76}$$

The constant C is given by the relation

$$C = \sigma_0 \int \left(k_0 + \frac{d\varphi_2(t)}{dt} \right) e^{[k_2 t + \varphi_2(t)]} \, dt \tag{2.77}$$

The following two typical limiting cases are discussed:

(a) If any further changes of strains are prevented immediately after the application of stress σ_0, i.e. if $t_r = t_0$, then from Eqs. (2.73) and (2.74), $k_0 = k_2 = k$. Substituting into Eq. (2.77) and integrating, we obtain

$$C = \sigma_0 e^{[kt + \varphi_2(t)]} + K \tag{2.78}$$

where K is a constant.

Substituting Eq. (2.78) into Eq. (2.76) and then into Eq. (2.69), we obtain, by considering the initial condition (Eq. (2.70)), the resulting relationship for the desired variation of stress relaxation

$$\sigma = \sigma_0 e^{-[k(t - t_r) + \varphi_2(t) - \varphi_2(t_r)]} \tag{2.79}$$

and, in accordance with Eq. (2.74)

$$\sigma = \sigma_0 e^{-[\varphi_1(t - t_r) - \varphi_1(0) + \varphi_2(t) - \varphi_2(t_r)]} \tag{2.80}$$

The latter expression acknowledges the effect of delayed elasticity and is a generalization of Eq. (2.66) which describes the variation of stress relaxation using the rate-of-creep theory.

(b) A more complex situation arises, if the instant t_0 at which the stress starts acting does not coincide with the instant t_r of the onset of the relaxation process (i.e. if $t_0 \neq t_r$). If this difference $t_r - t_0$ is large enough to assume that a substantial part of the delayed elastic strain arising from the stress component σ_0 has elapsed within the interval $t_r - t_0$, then it is possible to consider that $k_0 \approx 0$. Equation (2.77) is then simplified to the form

$$C = \sigma_0 \int \frac{d\varphi_2(t)}{dt} e^{[k_2 t + \varphi_2(t)]} \, dt \tag{2.82}$$

which, using the current functional expression for the creep coefficient, may be approximately replaced by the relation

$$C = \sigma_0 \int \left[P(t) + e^{\varphi_2(t)} \frac{d\varphi_2(t)}{dt} \right] dt \tag{2.83}$$

It has been found that the function $P(t)$ is approximately proportional to $\sqrt{t}$; after integration, it may be written

$$C = \sigma_0 [Q(t) + e^{\varphi_2(t)}] + K \tag{2.84}$$

84

where $Q(t) = \int P(t)\,dt$;

K = a constant.

Considering the initial conditions: Eq. (2.70) and Eq. (2.69), the resulting relationship for the stress relaxation, which occurs at a later period following the application of stress, is

$$\sigma(t) = \sigma_0 \{1 - \{[Q(t) - Q(t_r)] + e^{\varphi_2(t)} - e^{\varphi_2(t_r)}\} e^{-[k_2 t + \varphi_2(t)]}\} \tag{2.85}$$

The accuracy of this relationship has been proved by a comparison with the results of the numerical solution of an integral equation of the type: Eq. (2.22). It is interesting to compare both the described limiting cases with the analysis carried out by means of the rate-of-creep theory. While in the case (a), when relaxation started to develop directly from the instant of the load application, the decrease of stress (see Eq. (2.80)) is greater than that obtained via the rate-of-creep theory, the opposite result is observed in the case (b).

The described method of analysis allows the problem to be transformed into a simple differential equation of the type similar to that used in the analysis based on the rate-of-creep theory (Eq. (2.65)). The character of formulae respect, to some extent, the very significant effect of the delayed elasticity.

2.4.3   Analysis considering the general theory

As noted earlier, the analysis of relaxation with a general expression of the creep coefficient φ involves the solution of an integral equation of the Volterra type of the second order for the desired stress σ. In a practical computation, this analysis is carried out numerically.

2.4.3.1 *The time-discretization method on the basis of Eq. (2.20)*

This procedure corresponds to the calculation of deformations by means of summation with the use of partial time intervals within which the stress is constant (see Sect. 2.3.2). The simplest way is to subdivide the period of duration of the relaxation process (from its origin at time t_r, when variations of the strains are beginning to be restrained, until the investigated instant t) into N equal time intervals of length $\Delta t = \dfrac{t - t_r}{N}$ [*)]; at their centres τ_i, stresses $(-\Delta\sigma)$ start acting, which ensure that the strain at the end of each of these partial time intervals t_i

---

*) The computation is more efficient, if time intervals are used that are constant on a logarithm scale. This is usually necessary if long times must be reached in computation.

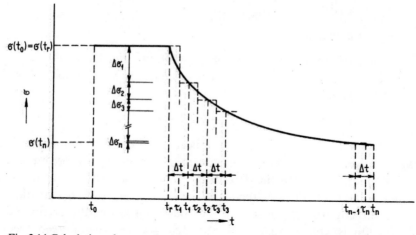

Fig. 2.14 Calculation of stress relaxation according to the time-discretization procedure

equals the strain at the beginning of relaxation [2.8]. This process is represented in Fig. 2.14.

At the beginning of relaxation, i.e. at time t_r (if E = constant is assumed, which, however, is not necessary when this method is applied), the strain is

$$\varepsilon(t_r) = \frac{1}{E}\{\sigma(t_0)[1 + \varphi(t_r, t_0)]\} \tag{2.86}$$

At the end of the first time interval, in accordance with Eq. (2.20), the strain is

$$\varepsilon(t_1) = \frac{1}{E}\{\sigma(t_0)[1 + \varphi(t_1, t_0)] - \Delta\sigma_1[1 + \varphi(t_1, \tau_1)]\} \tag{2.87}$$

Since, for relaxation

$$\varepsilon(t_1) = \varepsilon(t_r)$$

the following expression for the variation of the stress is obtained by comparing Eqs. (2.86) and (2.87)

$$\Delta\sigma_1 = \sigma(t_0)\frac{\varphi(t_1, t_0) - \varphi(t_r, t_0)}{1 + \varphi(t_1, \tau_1)} \tag{2.88}$$

analogously

$$\varepsilon(t_2) = \frac{1}{E}\{\sigma(t_0)[1 + \varphi(t_2, t_0)] - \Delta\sigma_1[1 + \varphi(t_2, \tau_1)] -$$
$$- \Delta\sigma_2[1 + \varphi(t_2, \tau_2)]\} \tag{2.89}$$

and, as before

$$\varepsilon(t_2) = \varepsilon(t_r)$$

Hence, comparing Eqs. (2.86) and (2.89):

$$\Delta\sigma_2 = \frac{\sigma(t_0)\left[\varphi(t_2, t_0) - \varphi(t_r, t_0)\right] - \Delta\sigma_1\left[1 + \varphi(t_2, \tau_1)\right]}{1 + \varphi(t_2, \tau_2)} \tag{2.90}$$

where $\Delta\sigma_1$ has been determined in the preceding interval using Eq. (2.88). This procedure may be repeated for all the time intervals. Generally, the recurrent formula for the k-th decrease of stress is

$$\Delta\sigma_k = \frac{\sigma(t_0)\left[\varphi(t_k, t_0) - \varphi(t_r, t_0)\right] - \sum_{i=1}^{k-1}\Delta\sigma_i\left[1 + \varphi(t_k, \tau_i)\right]}{1 + \varphi(t_k, \tau_k)} \tag{2.91}$$

where

$$t_k = t_r + k \cdot \Delta t$$
$$\tau_k = t_r + (k - 0.5)\,\Delta t \qquad k = 1, 2, \ldots, N$$

Then, the magnitude of stress at time t_k is

$$\sigma(t_k) = \sigma(t_0) - \sum_{i=1}^{k}\Delta\sigma_i \tag{2.92}$$

For computer programming, the expression in matrix form is convenient

$$[A]\{\Delta\sigma\} = \{r\}$$

where

$$[A] = \begin{bmatrix} 1 + \varphi(t_1, \tau_1), & 0, & \ldots, & 0 \\ 1 + \varphi(t_2, \tau_1), & 1 + \varphi(t_2, \tau_2), & \ldots, & 0 \\ \vdots & \vdots & & \vdots \\ 1 + \varphi(t_N, \tau_1), & 1 + \varphi(t_N, \tau_2), & \ldots, & 1 + \varphi(t_N, \tau_N) \end{bmatrix}$$

$$\{\Delta\sigma\} = \begin{Bmatrix} \Delta\sigma_1 \\ \Delta\sigma_2 \\ \vdots \\ \Delta\sigma_N \end{Bmatrix}, \quad \{r\} = \sigma(t_0) \cdot \begin{Bmatrix} \varphi(t_1, t) - \varphi(t_r, t_0) \\ \varphi(t_2, t_0) - \varphi(t_r, t_0) \\ \vdots \\ \varphi(t_N, t_0) - (t_r, t_0) \end{Bmatrix}$$

The calculation is very simple. First, the density of division N is chosen, then the values of the creep coefficient at the respective times t_k and τ_k are calculated, and the elements of the matrices $[A]$ and $\{r\}$ are determined. As the matrix $[A]$ contains non-zero elements only on the main diagonal and below it, $\Delta\sigma_1$ may be computed from the first equation, and then substituted into the second equation so that $\Delta\sigma_2$ may be determined, etc. The magnitude of stress after relaxation is determined from Eq. (2.92).

The accuracy of the analysis of relaxation increases with the number of the partial time intervals (with the density of division N). It has been proved by comparative calculations, however, that acceptable results are obtained even with a very small number of these time intervals. Thus, for example, in the case of a concrete structural member having $E = 38.5 \times 10^3$ MPa, cross-sectional area $A = 1.08 \times 10^6$ mm^2, cross-sectional perimeter $u = 4\,800$ mm, for $\varkappa_a = 3$, $\varphi_{fun} = 2$ (free space), subjected to a stress of 10 MPa at time $t_0 = 28$ days, at the starting time of the relaxation process $t_r = 180$ days and observing the stress at time $t = 1\,180$ days, we obtain, depending on the number of partial time intervals, the results listed in Table 2.1 and portrayed in Fig. 2.15.

Evidently, the values converge very rapidly. The difference between the ultimate values of stress $\sigma(t)$ for $N = 15$ and $N = 100$ is less than 0.09 per cent, and even for $N = 1$ it is only 9.1 per cent. Hence, for practical analyses a division into a small number of partial time intervals is adequate. For an error of less than 1 per cent, $N = 4$ is satisfactory in most cases.

However, if we compare relaxation resulting from some published data based on the rate-of-creep theory with relaxation obtained when the general form of creep function is used, we find considerable differences. Thus, for example, in the case analysed above (for $N = 100$) we obtain the results listed in Table 2.2 and graphically represented in Fig. 2.16. The computed stresses differ by as much as 25 per cent, which may be another argument for abandoning the rate-of-creep theory.

Table 2.1

Density of subdivision	1	2	4	8	10	15	40	100
Stress $\sigma(t)$ [MPa]	5.388	5.727	5.863	5.909	5.915	5.921	5.926	5.927

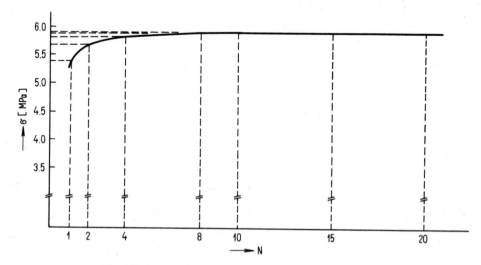

Fig. 2.15 Convergence rate of the time-discretization method

Table 2.2

t [days]	180	305	430	555	680	805	930	1 055	1 180
Rate-of-creep theory σ [MPa]	10	7.750	6.611	5.910	5.431	5.082	4.816	4.607	4.437
General theory σ [MPa]	10	8.448	7.694	7.158	6.800	6.500	6.278	6.081	5.927

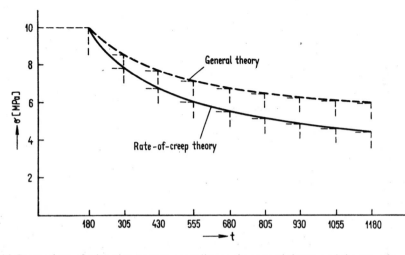

Fig. 2.16 Comparison of relaxation process according to the general theory and the rate-of-creep theory

It has been stated earlier that the rate-of-creep theory does not take into account the effect of previous loading. In the case of relaxation, this means that the effect of the instant of loading t_0 is not as influential as the beginning of relaxation t_r (see the exponent in Eq. (2.66)). The analyses using the general form of creep function, however, prove the significance of the effect of stress acting before the beginning of relaxation, i.e. within the interval $t_0 - t_r$. For the above specified concrete structural member, if the time t_0 of load application is varied, the values of the ultimate stresses (for $t = 1\ 180$ days) are rather different from each other (Table 2.3, Fig. 2.17).

The variation of the obtained results demonstrates also the effect of the delayed elasticity: if the time interval between the loading at instant t_0 and the beginning of relaxation t_r is large, some components of the strain due to the initial stress develop before the beginning of relaxation, and, vice versa, the decrements of stress arising from relaxation, having the character of reversed stresses (in our case, they are the stresses $\Delta\sigma$), manifest themselves upon the strain more.

Table 2.3

Age at loading t_0 [days]	28	56	90	120	150	180
General theory (σ) [MPa]	5.926	5.865	5.765	5.641	5.454	5.054
Rate-of-creep theory (σ) [MPa]	4.437	4.437	4.437	4.437	4.437	4.437
Error	25.13%					12.21%

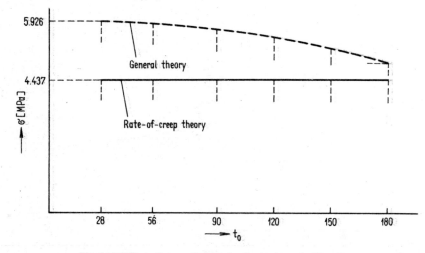

Fig. 2.17 Effect of stress before the beginning of relaxation

For this reason, smaller stress decrements are necessary for the maintenance of constant strain in the course of relaxation and the resulting stress has a higher value. Hence, for this case (a large time interval between t_0 and t_r) the manifestation of the relaxation is less than if concrete were loaded shortly before the beginning of relaxation (Fig. 2.17). Evidently, these facts are not regarded by the rate-of-creep theory.

Besides the length of the time interval $t_0 \ldots t_r$ (i.e. the period when the stress was acting before the time of imposition of constant strain), stress changes during this period also influence the stress variation caused by relaxation. The solution of this problem using the time-discretization method is again based on the condition that the strain at the ends of all time intervals equals the strain at the beginning of the relaxation process. If the simplest case is considered, when in addition to the stress $\sigma(t_0)$ imposed at time t_0 there is also acting a stress increment $\bar{\sigma}$ since time t_k (Fig. 2.18), the following formulae for stress decrements are obtained:

$$\Delta\sigma_1 = \sigma(t_0)\frac{\varphi(t_1, t_0) - \varphi(t_r, t_0)}{1 + \varphi(t_1, \tau_1)} + \bar{\sigma}\frac{\varphi(t_1, t_k) - \varphi(t_r, t_k)}{1 + \varphi(t_1, \tau_1)} \qquad (2.88)^*$$

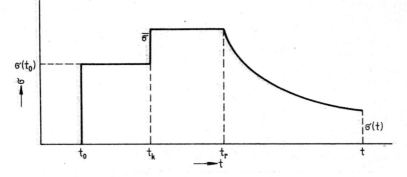

Fig. 2.18 Stress change before the beginning of relaxation

(which corresponds to Eq. 2.88)

$$\Delta\sigma_2 = \sigma(t_0) \frac{\varphi(t_2, t_0) - \varphi(t_r, t_0) - \dfrac{\varphi(t_1, t_0) - \varphi(t_r, t_0)}{1 + \varphi(t_1, \tau_1)} [1 + \varphi(t_2, \tau_1)]}{1 + \varphi(t_2, \tau_2)} +$$

$$+ \bar{\sigma} \frac{\varphi(t_2, t_k) - \varphi(t_r, t_k) - \dfrac{\varphi(t_1, t_k) - \varphi(t_r, t_k)}{1 + \varphi(t_1, \tau_1)} [1 + \varphi(t_2, \tau_1)]}{1 + \varphi(t_2, \tau_2)} \qquad (2.90)^*$$

(which corresponds to Eq. (2.93), etc.).

It is evident from the formulae (and follows also from the principle of super-position) that the effects of both stresses $\sigma(t_0)$ and $\bar{\sigma}$ run independently, and it is not possible to eliminate the influence of the intermediate time t_k.

The discussed time-discretization method is based on the condition that the magnitude of strain, beginning with instant t_r, does not vary any more (see, for example, Eqs. (2.89)). The procedure may be modified by using a fully equivalent condition stating that the increment of strain within the considered time interval is zero. Then we obtain successively:
For the first time interval

$$\Delta\varepsilon(t_1, t_r) = \sigma(t_0) [\varphi(t_1, t_0) - \varphi(t_r, t_0)] - \Delta\sigma_1 [1 + \varphi(t_1, \tau_1)] = 0$$

which yields

$$\Delta\sigma_1 = \sigma(t_0) \frac{\varphi(t_1, t_0) - \varphi(t_r, t_0)}{1 + \varphi(t_1, \tau_1)} \qquad (2.\overline{88})$$

which is completely in agreement with Eq. (2.88).
For the second time interval

$$\Delta\varepsilon(t_2, t_1) = \sigma(t_0) [\varphi(t_2, t_0) - \varphi(t_1, t_0)] - \Delta\sigma_1 [\varphi(t_2, \tau_1) - \varphi(t_1, \tau_1)] -$$
$$- \Delta\sigma_2 [1 + \varphi(t_2, \tau_2)] = 0$$

91

which yields the relation

$$\Delta\sigma_2 = \frac{\sigma(t_0)\left[\varphi(t_2, t_0) - \varphi(t_1, t_0)\right] - \Delta\sigma_1\left[\varphi(t_2, \tau_1) - \varphi(t_1, \tau_1)\right]}{1 + \varphi(t_2, \tau_2)} \qquad (2.\overline{90})$$

which corresponds to Eq. (2.90)* in this procedure, etc.

2.4.3.2 *Effective time method* [4.5]

It would be most convenient for practical use if the solution of relaxation problems could be undertaken with only a single time interval, i.e. without the need to subdivide the relaxation period. This is the basis of the effective time method.

The approximate method is similar to the general time-discretization method presented in the preceding Section 2.4.3.1 which has now degenerated to a single time interval, the length of which equals the whole relaxation period. The stress decrement $\Delta\sigma$, however, is not at the middle of the time interval as it was in the general time-discretization method, but at an appropriate *effective* time τ^* not known in advance (Fig. 2.19).

If this effective time were known, it would be easily possible to evaluate the desired stress decrement $\Delta\sigma$ analogously to Eq. (2.88):

$$\Delta\sigma = \sigma(t_0)\frac{\varphi(t, t_0) - \varphi(t_r, t_0)}{1 + \varphi(t, \tau^*)} \qquad (2.93)$$

If the rate-of-creep theory is applied, the stress decrement $\Delta\sigma$ can be (in accordance with Eqs. (2.66) and (2.92)) expressed as

$$\Delta\sigma = \sigma(t_0) - \sigma(t_r)\,e^{-[\varphi(t) - \varphi(t_r)]}$$

which, because $\sigma(t_0) = \sigma(t_r)$, can be written in the form

$$\Delta\sigma = \sigma(t_0)\left(1 - e^{-[\varphi(t) - \varphi(t_r)]}\right) \qquad (2.94)$$

Assuming the rate-of-creep theory, it follows that

$$\varphi(t, t_0) = \varphi(t) - \varphi(t_0)$$
$$\varphi(t_r, t_0) = \varphi(t_r) - \varphi(t_0)$$
$$\varphi(t, \tau^*) = \varphi(t) - \varphi(\tau^*)$$

---

*) The numerator in Eq. (2.$\overline{90}$) may be written in the form

$\sigma(t_0)\left[\varphi(t_2, t_0) - \varphi(t_1, t_0) - \varphi(t_r, t_0) + \varphi(t_r, t_0)\right] - \Delta\sigma_1\left[\varphi(t_2, \tau_1) - \varphi(t_1, \tau_1) + 1 - 1\right] =$

$= \sigma(t_0)\left[\varphi(t_2, t_0) - \varphi(t_r, t_0)\right] - \Delta\sigma_1\left[1 + \varphi(t_2, \tau_1)\right] - \underbrace{\sigma(t_0)\left[\varphi(t_1, t_0) - \varphi(t_r, t_0)\right] + \Delta\sigma_1\left[1 + \varphi(t_1, \tau_1)\right]}$

$\qquad\qquad\qquad\qquad\qquad\qquad\qquad\qquad\qquad\qquad\qquad 0 \quad \text{(see Eq. (2.88))}$

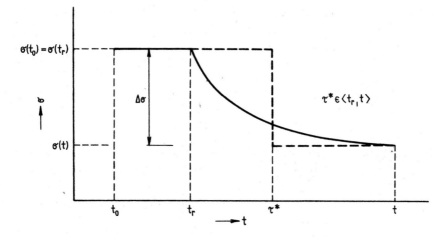

Fig. 2.19 Concept of the method of an effective time

Substituting these expressions into Eq. (2.93) and eliminating the stress decrement $\Delta\sigma$ (Eq. (2.94)), we obtain

$$1 - e^{-[\varphi(t) - \varphi(t_r)]} = \frac{\varphi(t) - \varphi(t_r)}{1 + \varphi(t) - \varphi(\tau^*)} \tag{2.95}$$

whence it follows

$$\varphi(\tau^*) = 1 + \varphi(t) - \frac{\varphi(t) - \varphi(t_r)}{1 - e^{-[\varphi(t) - \varphi(t_r)]}} \tag{2.96}$$

which defines the effective time τ^* corresponding to the application of the rate-of-creep theory. It has been found that, in spite of the many shortcomings of the rate-of-creep theory which make it unsuitable for analyses, the effective time τ^* is fully satisfactory even for an analysis based on the general expression of the creep coefficient (Sect. 2.1.3). As an example (Table 2.1 and Fig. 2.15), the effective time τ^* determined with reference to Eq. (2.96) is 405 days, and the value of stress after relaxation $\sigma(t) = 5.946$ MPa; the error is only 0.36 per cent of the accurate value (for $N = 100$).

For practical analysis, the following procedure is then applicable:

(i) For the given instants t_r and t, the values of the creep functions $\varphi(t_r)$ and $\varphi(t)$, respectively, are determined from the available Tables (for the rate-of-creep theory);

(ii) Substituting into Eq. (2.96), the value of the function $\varphi(\tau^*)$ is determined and, conversely, the required τ^* from the Tables.

(iii) The analysis using the effective time τ^* is carried out by way of the general theory of creep using the earlier procedure with a single interval, in accordance with Eq. (2.93).

The effective time τ^* may also be determined directly from (2.96), viz.

$$\tau^* = \left[\ln \frac{\varphi_\infty^2}{\varphi_\infty^2 - \left(1 + \varphi(t) - \dfrac{\varphi(t) - \varphi(t_r)}{1 - e^{-[\varphi(t) - \varphi(t_r)]}}\right)^2} \right]^2 \qquad (2.97)$$

where

$$\varphi(t) = \varphi_\infty \sqrt{1 - e^{-\sqrt{t}}}$$

In this way, the analysis of relaxation is transformed into a very simple relation given by Eq. (2.93), which is still simpler than the relation resulting from the rate-of-creep theory; at the same time, this formula is almost perfectly accurate.

The determination of the effective time can be simplified as follows: Eq. (2.96) may be written in the form

$$\varphi(\tau^*) = \varphi(t_r) + k_1 [\varphi(t) - \varphi(t_r)] \qquad (2.98)$$

where

$$k_1 = 1 + \frac{1}{\varphi(t) - \varphi(t_r)} - \frac{1}{1 - e^{-[\varphi(t) - \varphi(t_r)]}} =$$

$$= 1 + \frac{1}{\Delta\varphi} - \frac{1}{1 - e^{-\Delta\varphi}} \qquad (2.99)$$

In this way, the desired value of the creep coefficient (according to the rate-of-creep method) for the effective time τ^* is expressed as function of the coefficient $\varphi(t_r)$ at the beginning of relaxation, and the difference in values of the creep coefficients at the end and the beginning of the time interval, i.e. upon $\Delta\varphi = \varphi(t) - \varphi(t_r)$.

Equation (2.99) is basically independent on the form of expression of creep coefficient φ used. The coefficient k_1 plotted as a function of $\Delta\varphi$, in accordance with Eq. (2.99), is depicted in Fig. 2.20a by a solid line. The shape of this curve can be approximated by a straight line (the dashed line in Fig. 2.20a). Therefore, it is possible to write

$$k_1 \doteq 0.5 - 0.077 \, \Delta\varphi \qquad (2.100)$$

so that by using Eq. (2.98), $\varphi(\tau^*)$ can be determined.

It seems to be more convenient to determine the effective time τ^* directly, for example, by the formula

$$\tau^* = \frac{t + t_r}{k_2} \qquad (2.101)$$

This approach, however, is convenient only as long as the creep coefficient depends solely upon time. In this case, the coefficient k_2 could be taken from the diagrams whose example is shown in Fig. 2.20b. If we take into account that the

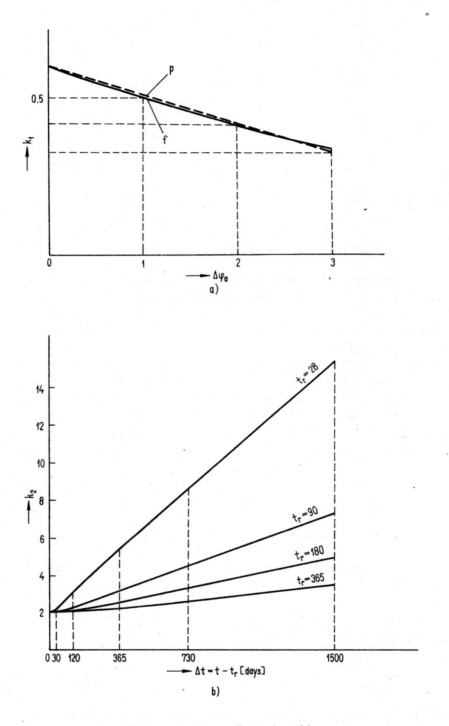

Fig. 2.20 Diagrams of the coefficients k_1 and k_2

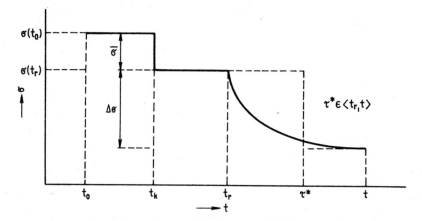

Fig. 2.21 Stress change before the beginning of relaxation — an effective time approach

creep characteristics depend on the humidity conditions and on the size or thickness of the cross-section, we would need different diagrams for evaluating the coefficient k_2. It would lead to difficulties and, therefore, the calculation based on Eq. (2.100) where the form of the creep coefficient need not be specified is usually the most convenient one.

Until now, relaxation was discussed for a particular case, in which a stress $\sigma(t_0)$ of given magnitude was introduced into a concrete structural member at time t_0, and this constant stress acted until the time of imposition of constant strain (Fig. 2.11). However, prior to the investigated time interval, an actual structure is subjected to a series of stress variations (for example, corresponding to the preceding states of the structure in the course of the construction process). For this reason, it is necessary to consider the loading history in more detail.

Let us consider the stress history portrayed in Fig. 2.21. At instant t_0, the concrete structural member is subjected to a stress $\sigma(t_0)$; at instant t_k, the stress $\sigma(t)$ decreases by a value $\bar{\sigma}$, and from instant t_r the relaxation process takes place. Hence, the magnitude of stress is $\sigma(t_r) = \sigma(t_0) - \bar{\sigma}$.

If the analysis of relaxation is carried out in one step with a decrease of stress $\Delta\sigma$ introduced at the effective time τ^* determined on the basis of Eq. (2.93), the following formula is obtained for the stress decrement due to relaxation (compare Eqs. (2.86) to (2.88)).

$$\Delta\sigma = \sigma(t_r)\frac{\varphi(t, t_0) - \varphi(t_r, t_0)}{1 + \varphi(t, \tau^*)} +$$

$$+ \bar{\sigma}\frac{\varphi(t, t_0) + \varphi(t_r, t_k) - \varphi(t, t_k) - \varphi(t_r, t_0)}{1 + \varphi(t, \tau^*)} \qquad (2.102)$$

which can be written in the form

$$\Delta\sigma = \Delta\sigma_r + \Delta\bar{\sigma} \qquad (2.103)$$

where $\Delta\sigma_r$ is the stress decrement for the case of a constant value of stress prior
to the commencement of the relaxation process;

$\Delta\bar{\sigma}$ expresses the effect of the change of stress prior to the commencement
of the relaxation process.

The effect of the change of stress prior to the onset of relaxation may be characterized,
with reference to Eq. (2.102), by the ratio

$$\frac{\Delta\bar{\sigma}}{\Delta\sigma_r} = \frac{\bar{\sigma}}{\sigma(t_r)}\left[1 - \frac{\varphi(t, t_k) - \varphi(t_r, t_k)}{\varphi(t, t_0) - \varphi(t_r, t_0)}\right] = \frac{\bar{\sigma}}{\sigma(t_r)} \cdot \alpha \qquad (2.104)$$

which is conveniently independent of the effective time τ^* at which the decrease of
stress $\Delta\sigma$ was introduced into the analysis.

Now, substituting $\Delta\bar{\sigma}$ from Eq. (2.104) into Eq. (2.103), we obtain the resulting
relationship for the evaluation of the decrement of stress

$$\Delta\sigma = \Delta\sigma_r\left[1 + \frac{\bar{\sigma}}{\sigma(t_r)} \cdot \alpha\right] \qquad (2.105)$$

Obviously, the coefficient α rather significantly affects the magnitude of $\Delta\sigma$.
It reaches its extreme value at $t_k = t_r$ ($\alpha = -0.256$ for the discussed example); at
$t_k \to t_0$ (i.e., reducing the time of action of the stress $\bar{\sigma}$), $\alpha \to 0$.

In case the stress $\sigma(t_0)$ increases by the value $\bar{\sigma}$ at instant t_k, the stress $\bar{\sigma}$ is
substituted into the presented relations with negative sign. If the magnitude of
stress σ varies several times within the time interval $\langle t_0, t_r \rangle$, the decrease of
stress $\Delta\sigma$ may be determined from the following relation representing a generalization of Eq. (2.105)

$$\Delta\sigma = \Delta\sigma_r\left[1 + \frac{1}{\sigma(t_r)} \cdot \sum_{k=1}^{n} \bar{\sigma}_k \alpha_k\right] \qquad (2.106)$$

It may be concluded that the analysis of relaxation by the methods introduced
in Sections 2.4.3.1 and 2.4.3.2 can be carried out in a relatively simple way, in spite
of the general function of creep prediction being used. In the current design
practice a convenient method of analysing relaxation seems to be that of effective
time carried out in one step; such an analysis may be completed without the aid
of a computer. For a more accurate determination of stress variations due to
relaxation, a computer-aided analysis according to the general method presented
in Sect. 2.4.3.1. is recommended.

2.4.3.3 *The time-discretization method based on Eq. (2.25)*

This procedure is the converse of that for to the calculation of strains described
in Sect. 2.3.3. As in the preceding method of analysis of relaxation, it is assumed
here that the stress is constant within the individual partial time intervals and that

it changes abruptly at the boundaries of these intervals. If we start from the method introduced in Sect. 2.3.3 for the calculation of strains with the given stress history, we may use Eqs. (2.44) and/or (2.45) for the analysis of the converse problem, i.e. for the analysis of stress variation caused by a prescribed constant strain. In these relations, the values of ε remain constant beginning with the onset of relaxation t_r, while the desired stress σ varies within this interval.

In the case of hand analysis, this means that, up to instant t_r, we proceed according to Eq. (2.44) (merely multiplying and adding the individual terms), then, in the course of the relaxation process, the corresponding values of stress are calculated from each subsequent equation. In terms of the triangular shape of the matrix $[F]$, this is a repeated calculation with always one unknown from one equation.

If the concrete structural member is loaded and if it is immediately restrained against further strains, i.e. if the strain remains equal to the value of the elastic strain ε_0, a special case occurs for which a particular analysis can be derived.

Since it holds that the i-th row of the matrix $[F]^{-1}$ (inverse matrix to the matrix $[F]$ in Eq. (2.45) — see also [2.5]), it is possible to write

$$- F_0 E_0 E_1(1 - F_1 E_1)(1 - F_2 E_2) \ldots (1 - F_{i-1} E_{i-1}), \ldots - F_{i-1} E_{i-1} E_i, E_i,$$
$$0 \ldots, 0 \tag{2.107}$$

In the case when the modulus of elasticity is constant, this relation may be simplified to

$$E(\varphi_1 - \varphi_0)(1 + \varphi_2 - \varphi_1) \ldots (1 + \varphi_i - \varphi_{i-1}),$$
$$E(\varphi_2 - \varphi_1)(1 + \varphi_3 - \varphi_2) \ldots (1 + \varphi_i - \varphi_{i-1}),$$
$$E(\varphi_i - \varphi_{i-1}), E, 0, 0 \ldots \tag{2.108}$$

The inverse relationship of Eq. (2.45) can then be written in the form

$$\sigma_k = E\varepsilon_k + E \sum_{j=0}^{k-1} H_j \varepsilon_j \tag{2.109}$$

where

$$H_j = (\varphi_{j+1} - \varphi_j)(1 + \varphi_{j+2} - \varphi_{j+1}) \ldots (1 + \varphi_k - \varphi_{k-1}) \tag{2.110}$$

When ε is constant, Eq. (2.109) acquires its final form

$$\sigma_k = E\varepsilon(1 + \sum_{j=0}^{k-1} H_j) = \tag{2.111}$$
$$= E\varepsilon(1 + \varphi_1 - \varphi_0)(1 + \varphi_2 - \varphi_1) \ldots (1 + \varphi_k - \varphi_{k-1})$$

2.4.3.4 *The analysis of relaxation with the use of a special form of the creep coefficient*

If the creep coefficient is expressed by means of Eq. (2.48), the analysis of relaxation is exceptionally easy. The calculation of strains for a given stress history, as stated in Sect. 2.3.4, leads to the simple relationship of Eq. (2.56) which determines the strain increment within the investigated time interval. The effect of the loading history is taken into account by means of easily applicable chain formulae (Eq. (2.57)) without the necessity of repeated calculations of the effects of all the previous loadings.

Hence, it suffices to consider the analysis of relaxation with the strain constant during the process and that, consequently, its increments are zero. Thus, for example, Eq. (2.56) reads

$$\Delta\sigma_3 + \sum_{i=1}^{m} A_{i4}(1 - a_{i4}) = 0 \tag{2.112}$$

which can be written, in accordance with Eq. (2.57),

$$\Delta\sigma_3 + \sum_{i=1}^{m} A_{i3}a_{i3}(1 - a_{i4}) + \Delta\sigma_3 \sum_{i=1}^{m} \bar{C}_i(t_3)(1 - a_{i4}) = 0 \tag{2.113}$$

Whence, the formula for the stress variation in consequence of relaxation is

$$\Delta\sigma_3 = \frac{-\sum_{i=1}^{m} A_{i3}a_{i3}(1 - a_{i4})}{1 + \sum_{i=1}^{m} \bar{C}_i(t_3)(1 - a_{i4})} \tag{2.114}$$

2.4.3.5 *Analysis with the use of the relaxation coefficient* [2.9]

Numerical techniques for the analysis of stress are laborious, but they usually represent the only possible approach to the determination of stress variations due to creep and shrinkage when the given relationships for the expression of the creep coefficients are complex. A simple approximate numerical analysis is possible using the relaxation coefficient. The following principles of this analysis are based on the form of creep function expressed in accordance with the CEB–FIP Recommendations (of 1970).

The creep coefficient is calculated from the expression

$$\varphi(t, \tau) = k_c k_b k_e k_d(\tau) k_t(t - \tau) = \bar{\varphi}_\infty k_d(\tau) k_t(t - \tau) \tag{2.115}$$

The strain due to shrinkage in the investigated period $(t - \tau)$ is given by

$$\varepsilon_s(t) = \varepsilon_{s\infty} k_t(t - \tau) \tag{2.116}$$

which can be written in the form

$$\varepsilon_s(t) = \frac{\varepsilon_{s\infty}}{\bar{\varphi}_\infty k_d(\tau)} \varphi(t, \tau) \tag{2.117}$$

The strain due to the prescribed stress history, expressed in terms of stress increments $\Delta\sigma(\tau)$ (Fig. 2.9), is given by Eq. (2.20). When considering the modulus of elasticity as constant in time and regarding the effect of shrinkage, the total strain may be expressed as

$$\varepsilon(t) = \frac{\sigma(t_0)}{E}[1 + \varphi(t, t_0)] + \sum_{i=1}^{N} \frac{\Delta\sigma_i(\tau_i)}{E}[1 + \varphi(t, \tau_i)] + \varepsilon_s(t) \tag{2.118}$$

in which $\sigma(t_0)$ is the first stress applied at time t_0. As the term $\sum_{i=1}^{N} \Delta\sigma_i(\tau_i) = \Delta\sigma(t)$ represents the total stress change, Eq. (2.118) takes the following form

$$\varepsilon(t) - \frac{\sigma(t_0)}{E} = \frac{\sigma(t_0)}{E} \varphi(t, t_0) + \varepsilon_s(t) + \frac{\Delta\sigma(t)}{E} +$$
$$+ \frac{\Delta\sigma_\infty}{E} \sum_{i=1}^{N} \frac{\Delta\sigma_i(\tau_i)}{\Delta\sigma_\infty}[\varphi(t, \tau_i)] \tag{2.119}$$

where $\Delta\sigma_\infty$ is the stress increment for $t \to \infty$, i.e.

$$\Delta\sigma_\infty = \sigma_\infty - \sigma(t_0) \tag{2.120}$$

Substitution of Eqs. (2.115) and (2.117) into Eq. (2.120) yields

$$\varepsilon(t) - \frac{\sigma(t_0)}{E} = \left[\frac{\sigma(t_0)}{E} + \frac{\varepsilon_{s\infty}}{\bar{\varphi}_\infty k_d(t_0)}\right] \bar{\varphi}_\infty k_d(t_0) k_t(t - t_0) +$$
$$+ \frac{\Delta\sigma(t)}{E} + \frac{\Delta\sigma_\infty}{E} \sum_{i=1}^{N} \frac{\Delta\sigma_i(\tau_i)}{\Delta\sigma_\infty} \bar{\varphi}_\infty k_d(\tau_i) k_t(t - \tau_i) \tag{2.121}$$

The initial strain is given by

$$\frac{\sigma(t_0)}{E} = \varepsilon(t_0) \tag{2.122}$$

Let us assume that the strain variation is governed by

$$\varepsilon(t) = \varepsilon(t_0) + [\varepsilon_\infty - \varepsilon(t_0)] k_t(t - t_0) \tag{2.123}$$

then Eq. (2.121) can be written in the form

$$E\{\varepsilon(t_0) + [\varepsilon_\infty - \varepsilon(t_0)] k_t(t - t_0) - \varepsilon(t_0) - \left[\sigma(t_0) + \frac{E\varepsilon_{s\infty}}{\bar{\varphi}_\infty k_d(t_0)}\right]$$
$$\bar{\varphi}_\infty k_d(t_0) k_t(t - t_0)\} = \Delta\sigma(t) + \Delta\sigma_\infty \bar{\varphi}_\infty k_d(t_0) \sum_{i=1}^{N} \frac{\Delta\sigma_i(\tau_i)}{\Delta\sigma_\infty} \frac{k_d(\tau_i)}{k_d(t_0)}$$
$$k_t(t - \tau_i) \tag{2.124}$$

which may be adjusted to the form

$$\Delta\sigma(t)\left[1 + \frac{\Delta\sigma_\infty}{\Delta\sigma(t)} \bar{\varphi}_\infty k_d(t_0) \sum_{i=1}^{N} \frac{\Delta\sigma_i(\tau_i)}{\Delta\sigma_\infty} \frac{k_d(\tau_i)}{k_d(t_0)} k_t(t - \tau_i)\right] =$$

$$= \left[E \frac{\varepsilon_\infty - \varepsilon(t_0)}{\bar{\varphi}_\infty k_d(t_0)} - \sigma(t_0) - \frac{E\varepsilon_{s\infty}}{\bar{\varphi}_\infty k_d(t_0)}\right] \bar{\varphi}_\infty k_d(t_0) k_t(t - t_0)$$

(2.125)

If the time interval is extended to $t \to \infty$, the assumption

$$\lim_{t \to \infty} k_t(t - t_0) = 1$$

is valid. Eq. (2.125) then becomes

$$\Delta\sigma_\infty\left[1 + \bar{\varphi}_\infty k_d(t_0) \sum_{i=1}^{N} \frac{\Delta\sigma_i(\tau_i)}{\Delta\sigma_\infty} \frac{k_d(\tau_i)}{k_d(t_0)}\right] =$$

$$= \left[E \frac{\varepsilon_\infty - \varepsilon(t_0)}{\bar{\varphi}_\infty k_d(t_0)} - \sigma(t_0) - \frac{E\varepsilon_{s\infty}}{\bar{\varphi}_\infty k_d(t_0)}\right] \bar{\varphi}_\infty k_d(t_0)$$

(2.126)

The expression

$$\bar{\varrho} = \sum_{i=1}^{N} \frac{\Delta\sigma_i(\tau_i)}{\Delta\sigma_\infty} \frac{k_d(\tau_i)}{k_d(t_0)} = \sum_{i=1}^{N} F(t) \frac{k_d(\tau_i)}{k_d(t_0)}$$

(2.127)

is called the relaxation coefficient. If its value is known, then it is possible, according to Eq. (2.126), to evaluate the stress change $\Delta\sigma_\infty = \sigma_\infty - \sigma(t_0)$. Referring to Eq. (2.120), it can be written

$$\sigma_\infty - \sigma(t_0) = -\frac{\bar{\varphi}_\infty k_d(t_0)}{1 + \bar{\varrho}\bar{\varphi}_\infty k_d(t_0)} \left\{\sigma(t_0) - \frac{E[\varepsilon_\infty - \varepsilon(t_0)]}{\bar{\varphi}_\infty k_d(t_0)} + \right.$$

$$\left. + \frac{E\varepsilon_{s\infty}}{\bar{\varphi}_\infty k_d(t_0)}\right\}$$

(2.128)

The magnitude of the relaxation coefficient cannot be calculated directly, because the ratio $\dfrac{\Delta\sigma(\tau)}{\Delta\sigma_\infty}$ is not known. However, the function

$$F(t) = \sum_{\tau=0}^{\tau \to \infty} \frac{\Delta\sigma(\tau)}{\Delta\sigma_\infty}$$

falls in the zone between the functions $F(t)$ corresponding to the rate-of-creep theory and to the theory of delayed elasticity (Sects. 2.1.1 and 2.1.2); these boundaries may be calculated, and it has been proved that the zone is very narrow, as may be observed, for example, in Fig. 2.22, where $k_{d0} = 1.2$, $\bar{\varphi}_\infty = 2.0$, and hence $k_{d0}\bar{\varphi}_x = 1.2 \times 2.0$. Consequently, it is possible to consider, for example, the least values and to calculate the relaxation coefficient for different values of $k_d(\tau)$; its value is a function of $\bar{\varphi}_\infty$ and $k_d(\tau)$ (Fig. 2.23).

For typical structural conditions, the value $\bar{\varrho}$ of the relaxation coefficients is normally within the interval $\langle 0.8, 0.9 \rangle$, and, with a continuous variation of stress

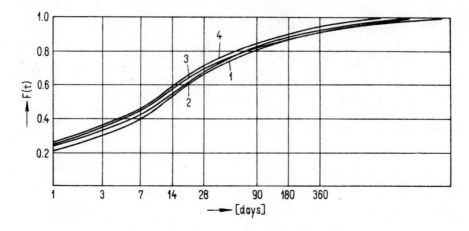

Fig. 2.22 Zone $F(t)$
(1) rate-of-creep theory; (2, 3) approximate solutions; (4) delayed elasticity theory

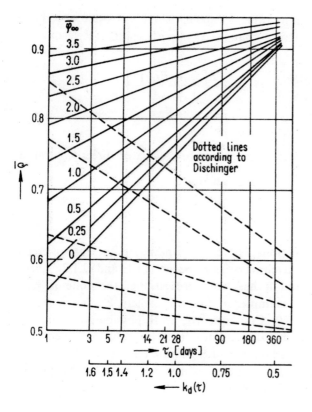

Fig. 2.23 Relaxation coefficient; τ_0 is
the age at the first loading

102

it may be assumed independent of time. Equation (2.128) can then be written in the form

$$\Delta\sigma(t) = \sigma(t) - \sigma(t_0) = - \frac{\varphi_\infty k_d(t_0) k_t(t - t_0)}{1 + \bar{\varrho}\bar{\varphi}_\infty k_d(t_0) k_t(t - t_0)}\Big[\sigma(t_0) -$$

$$- \frac{E[\varepsilon_\infty - \varepsilon(t_0)]}{\bar{\varphi}_\infty k_d(t_0)} + \frac{E\varepsilon_{s\infty}}{\bar{\varphi}_\infty k_d(t_0)}\Big] \tag{2.129}$$

2.4.3.6 *Effective modulus method*

The extreme simplification of the analysis of strain results from the assumption that the stress is constant in time; then the following applies

$$\varepsilon(t) = \frac{\sigma(t)}{E(\tau_1)}[1 + \varphi(t, \tau_1)] \tag{2.130}$$

However, this method can be only utilized in the case when the variation of the stress σ at time $\langle \tau_1, t \rangle$ is very small; for example, when the stress variations due to creep are calculated for nominally reinforced or prestressed cross-sections.

This in an old method consisting of a single elastic solution based on the effective modulus $E_r = E(\tau_1)/[1 + \varphi(t, \tau_1)]$. Accuracy is usually excellent when aging is negligible, as in very old concrete. In this case, the creep function depends only on $(t - \tau)$, i.e., creep curves for all τ are identical, but mutually displaced horizontally. This overestimates creep due to stress changes after τ_1 and incorrectly implies all creep to be totally recoverable after unloading [2.2].

2.4.3.7 *Age-adjusted effective modulus method* [2.2], [2.10]

As stated above, the simplest and the most widespread among the simplified methods of analysis is the effective modulus method (see Sect. 2.4.3.6), whose error with regard to the theoretically exact solution for the given creep law is known to be quite large with aging of concrete, i.e. when the change of its properties with the progress of hydration is of significance. Therefore, an improved method was originated by Trost in 1967 [2.9] (also see Sect. 2.4.3.5).

A refinement of this method with rigorous formulation has been presented by Bažant in [2.10] as an age-adjusted effective modulus method.

The change in stress values during a time period $t_0 \ldots t$ is expressed as

$$\Delta\sigma(t) = \sigma(t) - \sigma(t_0) = E''(t, t_0)\Big\{\varepsilon(t) - \varepsilon(t_0) - \frac{\sigma(t_0)}{E(t_0)}\varphi(t, t_0) -$$

$$- [\varepsilon_s(t) - \varepsilon_s(t_0)] - [\varepsilon_T(t) - \varepsilon_T(t_0)]\Big\} \tag{2.131}$$

103

in which ε_s and ε_T are the prescribed stress-independent inelastic strains representing shrinkage and thermal dilatations, $E''(t, t_0)$ is the age-adjusted effective modulus given by the following formula

$$E''(t, t_0) = \frac{E(t_0)}{1 + \chi(t, t_0) \, \varphi(t, t_0)} \tag{2.132}$$

where the aging coefficient $\chi(t, t_0)$ is expressed by [2.10]:

$$\chi(t, t_0) = \left[1 - \frac{R(t, t_0)}{E(t_0)} \right]^{-1} - \frac{1}{\varphi(t, t_0)} \tag{2.133}$$

$R(t, t_0)$ is the relaxation function = stress in time t caused by a unit strain introduced in time t_0.

Equation (2.131) has the form of Hooke's law and thus reduces the creep problem to a single elastic analysis, as in the usual effective modulus method. The values of E'' and χ have the same values for any strain history which is linear with the creep coefficient $(\varphi(t, t_0))$ and permits a sudden strain increment at the instant of the first loading. If the load changes instantly at times after the time of the first loading, the method can be also applied. The load history must then be considered as a sum of several step functions whose effects are analysed separately and finally superimposed.

Determination of the aging coefficient χ requires the knowledge of the relaxation function, which can be obtained from the creep function $J(t, \tau) = [1 + \varphi(t, \tau)]/E(\tau)$. This can be achieved by solving Volterra's integral equation, which is best carried out numerically [2.10].

As an example, Table 2.4 [2.10] shows the values of χ which have been found for the following creep law:

$$\varphi(t, \tau) = \varphi_u(\tau) (t - \tau)^{0.6}/[10 + (t - \tau)^{0.6}] \tag{2.134}$$

where

$$\varphi_u(\tau) = \varphi(\infty, 7)1.25\tau^{-0.118} \tag{2.135}$$

and τ is in days.

For a variable E, considered in Table 2.4, the following applies

$$E(\tau) = E(28) [\tau/(4 + 0.85\tau)]^{1/2} \tag{2.136}$$

It is evident from Table 2.4 [2.10] that the time-variation of E is quite significant, and its omission might be responsible for large errors.

The theoretical accuracy of the age-adjusted effective modulus method appears to be distinctly superior to that of the usual effective modulus method, while in simplicity both methods are equal. The method is also much more accurate than the rate-of-creep method.

Table 2.4 (according to [2.10])

Days $t - t_0$	$\varphi(\infty, 7)$	Variable E				Constant E				$\dfrac{\varphi(t, t_0)}{\varphi_u(t_0)}$
		t_0 [days]				t_0 [days]				
		10^1	10^2	10^3	10^4	10^1	10^2	10^3	10^4	
10^1	0.5	0.525	0.804	0.811	0.809	0.798	0.811	0.811	0.809	0.273
	1.5	0.720	0.826	0.825	0.820	0.820	0.829	0.825	0.820	
	2.5	0.774	0.842	0.837	0.830	0.839	0.844	0.837	0.830	
	3.5	0.806	0.856	0.848	0.839	0.855	0.857	0.848	0.839	
10^2	0.5	0.505	0.888	0.916	0.915	0.848	0.905	0.916	0.915	0.608
	1.5	0.739	0.919	0.932	0.928	0.878	0.926	0.932	0.928	
	2.5	0.804	0.935	0.943	0.938	0.899	0.939	0.943	0.938	
	3.5	0.839	0.946	0.951	0.946	0.914	0.949	0.951	0.946	
10^3	0.5	0.511	0.912	0.973	0.981	0.846	0.937	0.974	0.981	0.857
	1.5	0.732	0.943	0.981	0.985	0.878	0.953	0.981	0.985	
	2.5	0.795	0.956	0.985	0.988	0.899	0.963	0.985	0.988	
	3.5	0.830	0.964	0.987	0.990	0.914	0.969	0.987	0.990	
10^4	0.5	0.501	0.899	0.976	0.994	0.828	0.927	0.977	0.994	0.954
	1.5	0.717	0.934	0.983	0.995	0.863	0.945	0.983	0.995	
	2.5	0.781	0.949	0.986	0.996	0.887	0.956	0.987	0.996	
	3.5	0.818	0.958	0.989	0.997	0.903	0.963	0.989	0.997	

This approach is, however, convenient only as long as the creep functions for different humidities and specimen sizes, plotted as functions of $t - \tau$ for various t values, have the same time shapes, i.e. are mutually proportional. If we take into account that the creep curves actually have rather different time-characteristics, depending on the humidity conditions and the size or thickness of the cross-section, we would need a different table of the aging coefficient for each substantially different time-characteristic of the creep curve. This would lead to impractical large tables.

In order to avoid this inconvenience, an approximate empirical formula for the relaxation function has been developed in [2.11]. The formula is generally applicable, does not require advance tabulation of any coefficient, and works well for all conceivable time-characteristics of the creep function.

In the absence of aging, creep only depends on the load duration ξ and not on the age at load application τ. Thus, the absence of aging is characterized by the condition

$$J(\tau + \xi, \tau) = J(t, t - \xi)$$

because the load duration for each of these two creep functions is the same.

105

Therefore, the non-dimensional parameter [2.11]

$$\alpha_0 = \frac{J(\tau + \xi, \tau)}{J(t, t - \xi)} - 1 \tag{2.137}$$

vanishes if there is no aging. Because creep for a greater age at loading is smaller, we have $J(\tau + \xi, \tau) > J(t, t - \xi)$ if there is aging (assuming that $\tau + \xi < t$), and therefore α_0 is positive. This suggests that Eq. (2.137) may be considered as a characteristic of aging.

The value of ξ must be positive and must not exceed $t - t_0$, since the relaxation function must not depend on the J values for times that exceed $t_0 + (t - t_0) = t$. Similarly, the relaxation function must not depend on the J values for loading ages less than t_0, because the relaxation function represents a response to stress and there is no stress before time t_0 (see [2.11]).

In reference [2.11], the following approximation of the relaxation function is given

$$\tilde{R}(t, t_0) = \frac{1 - \Delta_0}{J(t, t_0)} - \frac{0.115\alpha_0}{J(t, t - 1)} \tag{2.138}$$

The coefficient 0.115 was determined by optimizing results calculated exactly (from the integral equation) for various typical functions $J(t, t_0)$, [2.11].

The optimum value of ξ has been found to be

$$\xi = \frac{1}{2}(t - t_0) \tag{2.139}$$

The coefficient Δ_0 introduces a relatively minor, age-independent correction that is generally less than 0.02 and may be neglected, i.e. $\Delta_0 \doteq 0$. More accurately, except for $t - t_0 < 1$ day, it is possible to use $\Delta_0 \doteq 0.008$. Still more accurately, a variable coefficient Δ_0 may be used, and the following formula has been recommended, [2.11]

$$\Delta_0 = 0.009 \left[\frac{J(t_0 + 1, t_0)}{J(29, 28)} \right]^2 \frac{J(t, t_0) - J(t_0 + 0.01, t_0)}{J(t, t_0) - 0.9J(t_0 + 0.01, t_0)} \tag{2.140}$$

This formula gives $\Delta_0 \doteq 0.009 \left[J(t_0 + 1, t_0)/J(29, 28) \right]^2$ for $t - t_0 > 3$ days, which is independent of load duration $t - t_0$ and depends only on the age of loading, t_0. Equation (2.140) brings about an improvement for short creep durations, such as $t - t_0 < 3$ days. For long-term values, Eq. (2.140) is unnecessary.

The proposed formula is useful especially for the quasi-elastic creep analysis by the age-adjusted effective modulus method. The formula makes the tables of the aging coefficient unnecessary, and, in fact, makes even the notion of this coefficient superfluous. Instead of calculating the age-adjusted effective modulus $E''(t, t_0)$ from the aging coefficient (Eq. (2.132)), it is now possible to evaluate it directly in terms of the relaxation function as approximated by Eq. (2.138), i.e.

$$E''(t, t_0) \doteq \frac{E(t_0) - \tilde{R}(t, t_0)}{\varphi(t, t_0)} = \frac{1 - J(t^*, t_0)\,\tilde{R}(t, t_0)}{J(t, t_0) - J(t^*, t_0)} \tag{2.141}$$

in which $t^* = t_0 + \Delta t$ where Δt is the time interval within which the load is initially applied; $E(t_0) = 1/J(t^*, t_0)$ is the corresponding elastic modulus; the value of Δt is usually between 0.01 day and 1 day and may be taken, e.g., as 0.1 day. It is found that the long-term values of $E''(t, t_0)$ are rather insensitive to the chosen value of Δt, [2.11].

It is important to note that the proposed formula for $\tilde{R}(t, t_0)$ does not involve the initial elastic modulus and does not require any short-term deformation for loads of less than 1-day duration. This feature is particularly welcome since improper matching of incompatible creep coefficient and elastic modulus has been responsible for much of the confusion and inaccuracies in creep analysis of concrete structures, [2.11].

2.4.3.8 *Simplification with the aid of mean value*

Assuming that the stress σ varies linearly within a time interval $\langle t_0, t \rangle$, the fundamental equation for the analysis of creep effects may be modified by using the theorem of the medium value:

$$\begin{aligned} \varepsilon(t) = {} & \frac{\sigma(t_0)}{E(t_0)} + \frac{\sigma(t) + \sigma(t_0)}{2E(t_0)}\,\varphi(t, t_0) + \\ & + \frac{\sigma(t) - \sigma(t_0)}{2}\left[\frac{1}{E(t)} + \frac{1}{E(t_0)}\right] \end{aligned} \tag{2.142}$$

In this way, the set of integral equations is transformed into a system of linear equations. Here, it has been assumed that also the modulus of elasticity is variable in time. With a constant E, i.e. $E = E(t) = E(\tau)$, the equation acquires the form

$$\varepsilon(t) = \frac{\sigma(t)}{E} + \frac{\sigma(t) + \sigma(t_0)}{2E}\,\varphi(t, t_0) \tag{2.143}$$

This solution may be used, without any serious impairment of the result, only in case the difference of stresses within the given time interval is not more than 30 per cent (for example, in the analysis of stress variations in reinforced-concrete cross-sections, or in the analysis of the losses of prestressing and also in the analysis of composite sections); if this difference is exceeded, the time interval must be divided into smaller steps, or another method must be applied. The technique of the analysis is demonstrated for the rate-of-creep theory on an example in Chapter 3.

2.4.3.9 *Graphical method*

The analysis of the changes of stress in a single cross-section composed of differently aging materials, may also be carried out by way of the graphical method presented in [2.12], [2.13], [2.14].

2.5 Effect of temperature

Structures are influenced by the effects of temperature, simultaneously with the manifestations of creep and shrinkage of concrete. In structural members, uniform and non-uniform variations of the temperature cause deformational changes and, together with stresses induced by creep and shrinkage, may produce extensive redistributions of internal forces.

The temperature of a freely situated structure, without additional sources of heat, follows the temperature of the ambient surroundings; the time-dependent distribution of the temperature of the structure in the course of a year may be expressed by the approximate relationship

$$T(t) = T_z + \beta \sin \left[2\pi(t - t_z) \right] \tag{2.144}$$

where T_z is the average temperature;

β is the temperature amplitude;

t_z is the age of concrete at the time when the temperature of the structure has reached the value T_z;

t is the age of concrete in years.

Equation (2.144) is graphically expressed in Fig. 2.24.

The strain induced by thermal dilatations is

$$\varepsilon_T(t) = \alpha T(t) = \alpha\{T_z + \beta \sin \left[2\pi(t - t_z) \right]\} \tag{2.145}$$

where α is the coefficient of thermal dilatation of concrete.

If a concrete member is subjected to a stress $\sigma(\tau)$ and if shrinkage of concrete develops while the temperature varies, the strain of the member, with reference to Eq. (2.22), is described by the equation

$$\varepsilon(t) = \frac{\sigma(t_0)}{E(t_0)} [1 + \varphi(t, t_0)] + \int_{t_0}^{t} \frac{d\sigma(\tau)}{d\tau} \left[\frac{1}{E(\tau)} + \frac{\varphi(t, \tau)}{E(\tau)} \right] d\tau +$$

$$+ \varepsilon_s(t) + \alpha\{T_z + \beta \sin \left[2\pi(t - t_z) \right]\} \tag{2.146}$$

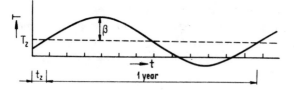

Fig. 2.24 Temperature variation in the course of a year

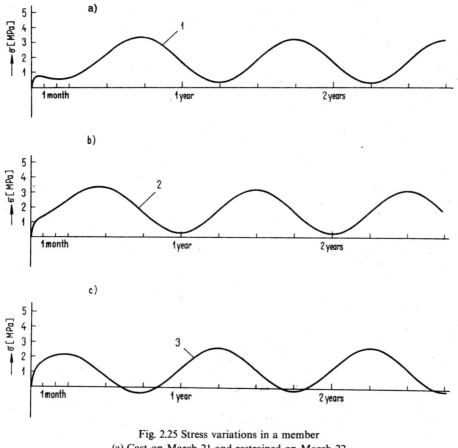

Fig. 2.25 Stress variations in a member
(a) Cast on March 21 and restrained on March 22
(b) Cast on June 21 and restrained on June 22
(c) Cast on September 23 and restrained on September 24

The analysis of deformations is simple: it is convenient to use the procedures explained in Sect. 2.3. The converse problem, i.e. the analysis of stress at a prescribed time-dependent variation of strain and temperature, is much more complicated. The procedures derived for the analysis of relaxation, introduced in Sect. 2.4, are suitable for the latter problem.

Let us investigate, for example, a structural member whose strains are restrained from the instant of casting the concrete or, in the case of precast elements, from the instant of erection. The thermal coefficient of expansion $\alpha = 0.000\,012$ and the amplitude of the temperature curve $\beta = 7.5\,°C$.

The stress history of the member of a cast-in-situ structure at various times of casting is portrayed in Fig. 2.25. The graph indicates that, in a concrete cast during the spring or summer season (Fig. 2.25a, b), the deformations induced by

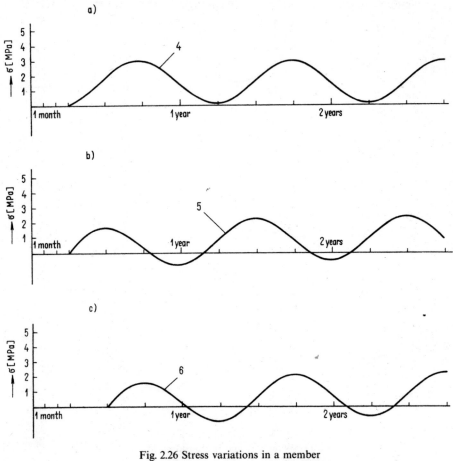

Fig. 2.26 Stress variations in a member
(a) Cast on March 21 and restrained on June 21
(b) Cast on June 21 and restrained on September 23
(c) Cast on March 21 and restrained on September 23

shrinkage are partly balanced by the expansion due to a higher summer temperature; because of this, a favourable reduction caused by the creep of concrete cannot take place. Consequently, large tensile stresses develop in the next winter season. These stresses also appear with nearly unchanged magnitudes in the following winters, because the effect of creep for an older concrete is much less.

On the other hand, in a concrete cast during the autumn season (Fig. 2.25c), when the effects of shrinkage and low temperature are added and reduced by the creep of young concrete, the tensile stresses in the winter seasons are much smaller and increase insignificantly in the course of the following years.

The stress history of a precast element of an erected structure is illustrated in Fig. 2.26.

110

The graph shows that it is rather disadvantageous to assemble a precast element aged 3 months during the summer season: although a part of shrinkage has already developed prior to erection, the low temperature after erection induces large tensile stresses in the element, and the difference between a cast-in-situ structure and one erected during the summer season is rather small (Fig. 2.27).

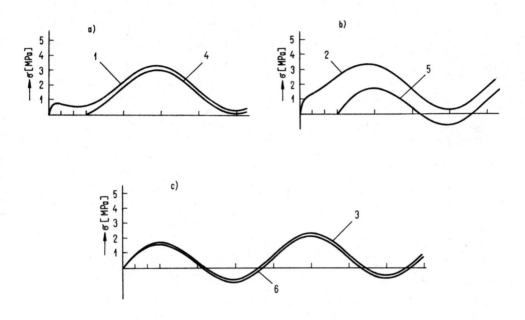

Fig. 2.27 Significance of the time interval at the
instant of making the structure monolithic

A more favourable state of the structure is established when it is erected during the autumn season, for a member precast in the spring (Fig. 2.27b) or in the summer (Fig. 2.27c). The tensile stresses are greatly reduced — even compressive stresses develop during the summer season, the advantage is evident from the comparison shown in Fig. 2.27b. On the other hand, the effect of the difference of age at the time of erection (3 months and 6 months) on the development of further stresses in the structure is not large (Fig. 2.27a).

The results of the above comparison are in agreement with the idea of an attenuation of the effects of shrinkage due to the creep of concrete, and they confirm the high significance of the effect of temperature on the stress developed in cast-in-situ and erected structures.

2.6 The approach of several Standards and Recommendations for creep and shrinkage

In this Chapter, the main principles concerning the effects of creep and shrinkage are presented and, from this point of view, four Recommendations are discussed: (i) the Recommendations of CEB–FIP, from the year 1970 [2.15]; (ii) the CEB–FIP Recommendations from 1978 [2.26]; (iii) the Soviet Recommendations SN 200–62/60, and (iiii) the Recommendations based on the ACI.

2.6.1 The Recommendations of CEB-FIP (1970) [2.15]

Creep

The strain is calculated by means of the expression

$$\varepsilon_c(t, t_0) = \frac{\sigma_0}{E_c} \varphi(t, t_0)$$

where σ_0 is the stress;

E_c is the modulus of elasticity at the age of 28 days;

$$\varphi(t) = k_c k_d k_b k_e k_t (t - t_0) \tag{2.147}$$

The coefficients k have the following meaning:

k_c expresses the influence of ambient conditions (Fig. 2.28);

k_d depends on the age of concrete (or on the degree of hardening) when the member was subjected to a load (Fig. 2.29) and on the type of cement; curve a relates to normal-hardening cement and curve b to cement with a high initial strength;

k_b depends on the composition of concrete (Fig. 2.30);

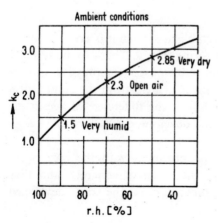

Fig. 2.28 Effect of ambient conditions

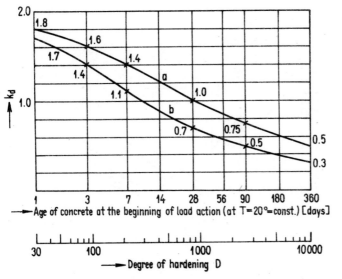

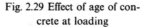

Fig. 2.29 Effect of age of concrete at loading

k_e expresses the influence of the fictitious depth of the member (Fig. 2.31);
k_t expresses the development of creep with time (Fig. 2.32).

The Recommendations stipulate that the value of the creep coefficient φ, calculated by means of the coefficients k should be considered as the mean value. Thus, if the creep effect could be critical for a limit state (e.g. serviceability), it is recommended to raise or to lower the calculated value of the coefficient φ by 15% in order to establish the most unfavourable case.

The values in Fig. 2.29 are valid for concretes on the basis of Portland cement that is hardened under normal conditions, i.e. at 20 °C temperature, and is protected against high losses of moisture. If the concrete hardens at other temperatures, the degree of hardening D should be taken as

$$D = \Sigma \, \Delta_t(T + 10)$$

D being the degree of hardening at the time of loading, Δ_t the number of days of hardening of the concrete at a temperature of T °C.

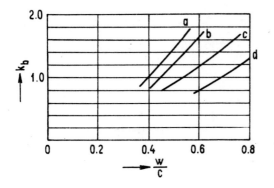

Fig. 2.30 Effect of amount of cement in concrete and of water/cement ratio
(a) 500 kg/m³; (b) 400 kg/m³; (c) 300 kg/m³; (d) 200 kg/m³

113

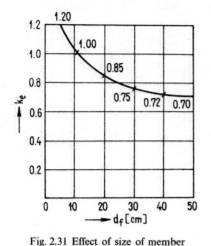

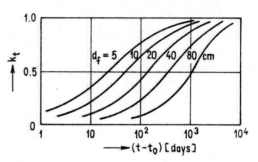

Fig. 2.31 Effect of size of member   Fig. 2.32 Relation between creep and time $t - t_0$

The fictitious depth is

$$d_f = \frac{2A_c}{U_c}$$

where A_c and U_c are the area and circumference of the member cross-section, respectively.

The coefficient k_t is calculated by means of the time period when the stress in the concrete began to act until the instant t:

$$k_t = k(t - t_0)$$

Shrinkage

The recommended formula is

$$\varepsilon_s = \varepsilon_{s0} k_b k_e' k_p k_t \tag{2.148}$$

where ε_{s0} expresses the influence of the ambient conditions on plain concrete (Fig. 2.33);

k_e' depends on the fictitious depth of the member (Fig. 2.34);

k_p depends on the steel ratio $\mu = \dfrac{A_s}{A_c} 100$ and is calculated by means of the expression

$$k_p = \frac{100}{100 + n\mu}$$

A_s and A_c are the areas of steel and concrete, respectively;
n (the ratio of the moduli of elasticity) $= 20$.
The coefficients k_b and k_t are the same as in the case of creep.

114

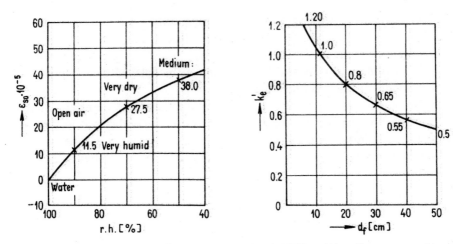

Fig. 2.33 Effect of ambient conditions on shrinkage

Fig. 2.34 Effect of size of concrete member on shrinkage

If the ambient conditions vary in time intervals, the part of shrinkage $\Delta\varepsilon_s(\tau - i)$ related to the time interval $(\tau - i)$ may be calculated using the relation

$$\Delta\varepsilon_s(\tau - i) = \varepsilon_{s0}k_bk'_ek_p(k_{t\tau} - k_{ti})$$

which means that the functions k_t are subtracted in the same way as in the rate-of-creep method.

2.6.2 The Recommendations of CEB–FIP (1978)

Creep [2.16]

The expression for the calculation of the strain is

$$\varepsilon_c(t, t_0) = \frac{\sigma_0}{E_{c28}} \varphi(t, t_0)$$

The coefficient $\varphi(t, t_0)$ has the form

$$\varphi(t, t_0) = \beta_a(t_0) + \varphi_d\beta_d(t - t_0) + \varphi_f[\beta_f(t) - \beta_f(t_0)] \qquad (2.149)$$

with

$$\beta_a(t_0) = 0.8\left(1 - \frac{R_c(t_0)}{R_c(\infty)}\right)$$

φ_d is a coefficient representing the delayed elasticity, its value being 0.4;

$\varphi_f = \varphi_{f1} \cdot \varphi_{f2}$ is a coefficient of delayed plasticity;

115

φ_{f1} is a coefficient depending on the ambient conditions;

φ_{f2} is a coefficient depending on the fictitious depth d_f;

β_d is a function standing for the development of the delayed elasticity with time;

β_f is a function expressing the development of the delayed plasticity with time; it depends on the fictitious depth;

t expresses the age of concrete at the time when the effect of creep is calculated;

t_0 expresses the age of concrete at the time when the member was loaded; both t and t_0 must be corrected with regard to temperature and the type of concrete.

$R_c(t_0)$ and $R_c(\infty)$ are the compressive strengths of concretes at times t_0 and $t \to \infty$, respectively.

The fictitious depth is calculated from the expression

$$d_f = \lambda \frac{2A_c}{U_c} \quad [\text{mm}]$$

If the ambient temperature of the hardening concrete differs greatly from 20° C, the time should be corrected with regard to temperature and type of cement: every actual period Δt_m, in which the medium ambient temperature is $T(t_m)$, should be converted into a fictitious period Δt by means of the relation

$$\Delta t = \frac{\alpha}{30} \sum_0^m \{[T(t_m) + 10] \Delta t_m\} \tag{2.150}$$

where α is a coefficient with the value of

1 for normal slow-hardening cement;

2 for rapid-hardening cement;

3 for rapid-hardening high-strength cement.

The values of the coefficients used in Eq. (2.149) are contained in Figs. 2.35 to 2.38 and in Table 2.5. The values of φ_{f1} and ε_{s1} listed in this Table are related to concretes with plastic consistency and they should be raised or lowered by 25 % when the concrete has a lower or higher consistency.

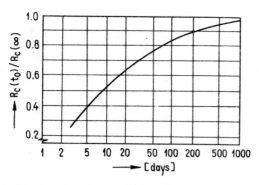

Fig. 2.35 Variation of strength of concrete in relation to age

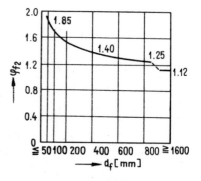

Fig. 2.36 Creep in relation to fictitious depth

116

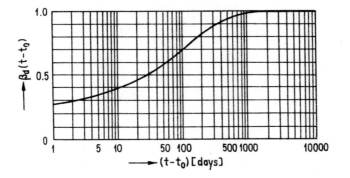

Fig. 2.37 Development of delayed elasticity with time

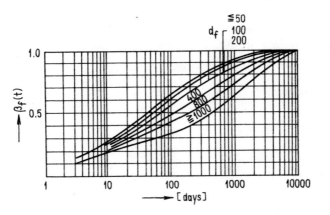

Fig. 2.38 Development of plastic deformation with time

Table 2.5

Ambient conditions	Relative humidity [%]	Coefficients		Coefficient λ
		creep φ_{f1}	shrinkage ε_{s1}	
1	2	3	4	5
water	—	0.8	+0.000 10	30
very humid	90	1.0	−0.000 13	5
in the open air	70	2.0	−0.000 32	1.5
very dry	40	3.0	−0.000 52	1

117

Basically, this method corresponds to the rate-of-creep method with delayed elasticity.

Shrinkage

The strain produced by shrinkage during the time interval $(t - \tau)$ is calculated by means of the formula

$$\varepsilon_s(t, \tau) = \varepsilon_{s0}[\beta_s(t) - \beta_s(\tau)] \tag{2.151}$$

where
$\varepsilon_{s0} = \varepsilon_{s1}\varepsilon_{s2}$ with
ε_{s1} dependent on the ambient conditions (Table 2.5),
ε_{s2} dependent on the fictitious depth d_f (Fig. 2.39);
β_s is a function of time (Fig. 2.40) and of the fictitious depth d_f;
t and τ are the ages of concrete (corrected with regard to Eq. (2.150) with $\alpha = 1$) when the effect of shrinkage is calculated and at the time from which the effect is considered, respectively.

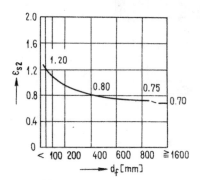

Fig. 2.39 Shrinkage related to fictitious depth

The functional expression of creep and shrinkage [2.7]

All the necessary quantities introduced above are given in graphs or numerically. As it is more practical to use analytical formulae in current calculations, the above coefficients can be expressed analytically as follows:

For creep

$$\beta_a(t_0) = 0.95t_0^{-0.3} - 0.1$$

$$\beta_d(t - t_0) = 0.28 + 0.5 \arctan [0.011(t - t_0)]^{\frac{2}{3}}$$

$$\varphi_{f2} = 1.108\,9 + 0.846\,9\,e^{-0.002\,7d_f}$$

$$\beta_f(t) = 1 - e^{-(A_f t)^{B_f}}; \quad \beta_f(t_0) = 1 - e^{-(A_f t_0)^{B_f}}$$

118

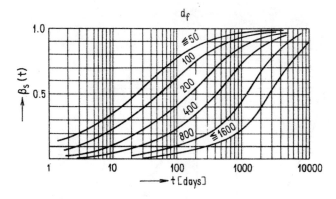

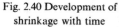

Fig. 2.40 Development of shrinkage with time

with

$$A_f = 0.078 \, e^{-1.22 \log d_f}$$

and

$$B_f = 0.528 \, e^{-0.13 \log d_f}$$

For shrinkage

$$\varepsilon_{s2} = 0.700 \, 1 + 0.629 \, 2 \, e^{-0.0046 d_f}$$

$$\beta_s(t) = 1 - e^{-(A_s t)^{B_s}}; \qquad \beta_s(\tau) = 1 - e^{-(A_s \tau)^{B_s}}$$

with

$$A_s = 3.16 \, e^{-2.98 \log d_f}$$

and

$$B_s = 0.18 \, e^{0.49 \log d_f}$$

The coefficients φ_{f1} and ε_{s1} and the calculation of creep and shrinkage are the same as explained earlier.

The differences between the values of the coefficients given both graphically and analytically are shown in Figs. 2.41 and 2.42 and in Tables 2.6a to 2.6d where the values of the coefficients with bars are analytical. Evidently, the functional formulae express the graphical relations fairly well.

Table 2.6a

τ [days]	3	5	10	20	50	100	200	500	1 000
$\beta_a(\tau)$	0.58	0.50	0.38	0.29	0.18	0.14	0.08	0.04	0.02
$\bar{\beta}_a(\tau)$	0.58	0.49	0.38	0.29	0.19	0.14	0.09	0.05	0.02

119

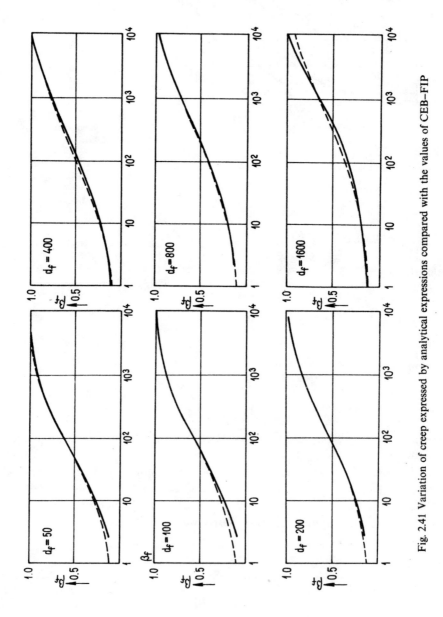

Fig. 2.41 Variation of creep expressed by analytical expressions compared with the values of CEB–FIP

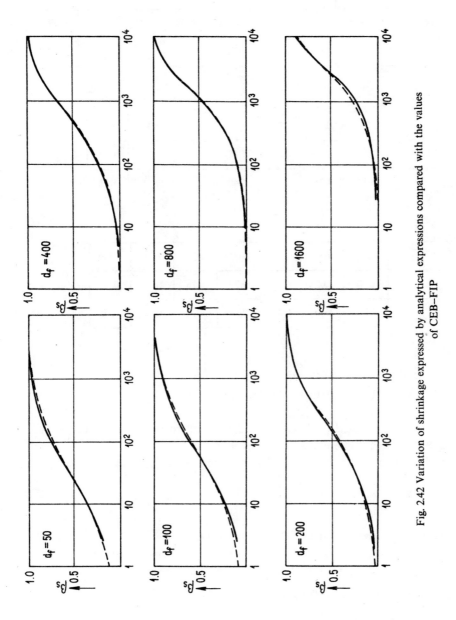

Fig. 2.42 Variation of shrinkage expressed by analytical expressions compared with the values of CEB–FIP

Table 2.6b

$(t - \tau)$ [days]	1	3	5	10	50	100	500	1 000	1 500
$\beta_d(t - \tau)$	0.28	0.32	0.35	0.40	0.58	0.69	0.93	0.99	1.00
$\bar{\beta}_d(t - \tau)$	0.30	0.33	0.35	0.39	0.58	0.69	0.91	0.97	0.99

Table 2.6c

d_f [mm]	50	100	200	400	600	800	1 600
φ_{f2}	1.85	1.70	1.55	1.40	1.30	1.25	1.12
$\bar{\varphi}_{f2}$	1.85	1.76	1.60	1.40	1.28	1.21	1.12

Table 2.6d

d_f [mm]	50	100	200	400	600	800	1 600
ε_{s2}	1.20	1.05	0.90	0.80	0.78	0.75	0.70
$\bar{\varepsilon}_{s2}$	1.20	1.10	0.95	0.80	0.74	0.72	0.70

2.6.3 Soviet Recommendations [2.17], [2.20] (with the original notation)

The results presented in this Section are taken from [2.17], where the principles of the Soviet Recommendations are explained.

The formulae for calculating the final creep coefficient and final shrinkage are

$$\varphi_\infty = \varphi^c_\infty \eta_1 \eta_2 \eta_3$$

$$\varepsilon_{s\infty} = \alpha^c \eta_1 \eta_2 \eta_4$$

These coefficients are listed in Tables 2.7 to 2.11.

The creep coefficient and the strain caused by shrinkage at an arbitrary time t may be calculated by means of the formulae

$$\varphi(t) = \varphi_\infty [1 - e^{-B(t - t_0)}]$$

$$\varepsilon_s(t) = \varepsilon_{s\infty} [1 - e^{-Bt}]$$

with

$$B = 0.04$$

Table 2.7

Concrete	φ^c_∞	$\alpha^c \times 10^5$
normal with coarse aggregate, normal-hardening	2.0	30
normal with coarse aggregate, steam-cured	1.8	25
lightweight with coarse aggregate, normal-hardening	2.5	35
normal with fine aggregate, normal-hardening	2.5	50
lightweight with fine aggregate, normal-hardening	2.5	50
lightweight with fine aggregate, steam-cured	2.0	40

Table 2.8   Coefficients μ_1 (influence of ambient conditions)

Ambient conditions [r.h. %]	μ_1	Ambient conditions [r.h. %]	μ_1
very dry (< 20) dry (20 ÷ 40) normal (50 ÷ 60)	1.40 1.30 1.00	humid (61 ÷ 75) wet (> 75) water	0.85 0.70 0.00

Table 2.9

d_{min}	μ_2	d_{min}	μ_2	d_{min}	μ_2
< 5	1.60	20	1.00	60	0.70
5	1.50	25	0.95	80	0.55
7	1.30	30	0.90	100	0.50
10	1.15	40	0.80	> 100	0.40
15	1.05	50	0.75		

d_{min} is the minimum dimension of the cross-section in cm

Table 2.10 Coefficients μ_3 (influence of the age of concrete at loading)

Age of concrete at the time of loading	3	5	7	10	14	20	28	40	60	90	180	360 days
μ_3	2.0 1.5	1.8 1.4	1.6 1.3	1.4 1.25	1.2	1.1	1.0	0.8	0.7	0.6	0.5	0.45

The top values relate to normal-hardening concrete.
The bottom values relate to steam- and heat-cured concrete.

Table 2.11 Coefficient μ_4 (influence of the age of concrete for shrinkage)

Time from which shrinkage is calculated [days]	$0 \div 3$	7	14	28	60	90	180	360
μ_4	1.00	0.95	0.75	0.55	0.35	0.30	0.25	0.20

2.6.4 Recommendations of the USA [2.18] (with the original notation)

The ACI Model Recommendations use the following expressions for creep

$$J(t, t') = \frac{1}{E(t')}\left(1 + \frac{(t - t')^{0.6}}{10 + (t - t')^{0.6}} C_u\right)$$

for shrinkage

$$\varepsilon_s(t, t_0) = \frac{t - t_0}{f_c + (t - t_0)} \varepsilon_u^s$$

where t', t and t_0 are the ages at loading in days, at the instant of observation and at the end of curing, respectively.

The elastic modulus is related to strength as follows:

$$E(t') = 33 \sqrt{\varrho^3 f_c'(t')}; \qquad f_c'(t') = f_{c\,28}' \frac{t'}{4 + 0.85 t'}$$

where ϱ is the unit weight of concrete (lb/cu ft) and $f_c'(t')$ is the cylinder strength of concrete (psi).

C_u equals 2.35, when the slump is 4 in or less, when the ambient relative humidity is 40 %, when the minimum dimension of the member is 6 in or less, and when the concrete was loaded at 7 days if moist-cured, and at 1 to 3 days if steam-cured.

For other cases

$$C_u = 2.35 \, K_t^c \, K_H^c \, K_T^c \, K_S^c \, K_F^c \, K_A^c$$

where

$$K_t^c = 1.25 t'^{-0.118} \qquad \text{for moist-cured concrete}$$
$$ 1.13 t'^{-0.095} \qquad \text{for steam-cured concrete}$$
$$K_H^c = 1.27 - 0.0067 \, h_e \qquad \text{where } h_e \geq 40\,\%$$
$$K_T^c = 1{:}14 - 0.023 \, T_m \qquad \text{for} \leq \text{one year loading}$$
$$ 1.10 - 0.017 \, T_m \qquad \text{for ultimate value}$$
$$K_S^c = 0.82 + 0.067 \, S_c$$

$$K_F^c = 0.88 + 0.002\,4\,F_a$$

$$
\begin{aligned}
K_A^c &= 1.00 & &\text{for } A_c \leq 6\% \\
&\quad 0.46 + 0.090\,A_c & &\text{for } A_c > 6\%
\end{aligned}
$$

t' is the age of concrete at loading (in days);

h_e is the ambient relative humidity in %;

T_m is the minimum dimension of the member in inches;

S_c is the slump in inches;

F_a is the percentage of fine aggregate by weight;

A_c is the air content in per cent of the concrete volume.

The ultimate shrinkage coefficient is specified as:

ε_u^s is $0.000\,800\,K_H^s K_T^s K_S^s K_B^s K_F^s K_A^s$ for moist-cured concrete, and $0.000\,730$ $K_H^s K_T^s K_S^s K_B^s K_F^s K_A^s$ for steam-cured concrete, where K_H^s, K_T^s, K_S^s, K_B^s, K_F^s and K_A^s are shrinkage correction factors. They equal 1.0 when the concrete has a slump of 4 in or less, when the ambient humidity is 40% and when the minimum thickness of the member is 6 in or less. In other cases, these factors are given by:

$$
\begin{aligned}
K_H^s &= 1.40 - 0.010\,h_e & &40\% \leq h_e \leq 80\% \\
K_H^s &= 3.00 - 0.030\,h_e & &80\% \leq h_e \leq 100\% \\
K_T^s &= 1.23 - 0.038\,T_m & &\text{for } \leq \text{ one year loading} \\
K_T^s &= 1.17 - 0.029\,T_m & &\text{for ultimate value} \\
K_S^s &= 0.89 + 0.041\,S_c \\
K_B^s &= 0.75 + 0.034\,B_s \\
K_F^s &= 0.30 + 0.014\,0\,F_a & &\text{for } F_a \leq 50\% \\
K_F^s &= 0.90 + 0.002\,0\,F_a & &\text{for } F_a \geq 50\% \\
K_A^s &= 0.95 + 0.008\,0\,A_c
\end{aligned}
$$

B_s is the number of 94-lb sacks of cement per one cubic yard of concrete.

Concerning f_c and t_0, the following values are recommended:

$f_c = 35$ days, $t_0 = 7$ days for moist-cured concrete;

$f_c = 55$ days, $t_0 = 1$ to 3 days for steam-cured concrete.

2.6.5 The Bažant–Panula Model [2.19]
(with the original notation)

This recent and very advanced model for the prediction of creep and shrinkage is defined as follows:

Shrinkage strain is given by:

$$\varepsilon_s(\hat{t}, t_0) = \varepsilon_{s\alpha}\,k_h S(\hat{t})$$

where t, t_0 and $\hat{t}$ are the ages of concrete, the age of concrete when drying begins and the duration of drying ($\hat{t} = t - t_0$), respectively.

$$S(\hat{t}) = \left(\frac{\hat{t}}{\tau_s + \hat{t}}\right)^{1/2}; \quad \tau_s = 600\left(\frac{k_s}{150}D\right)^2\frac{10}{C_1(t_0)}; \quad D = 2\frac{v}{s}$$

where v/s is the volume-to-surface ratio

$k_h = 1 - h^3$ for $h \leq 0.98$
$k_h = -0.2$ for $h = 1.0$,

h being the ambient relative humidity ($0 \leq h \leq 1$).

$$C_1(t) = C_7 k'_T(0.05 + \sqrt{6.3/t})$$

$$\varepsilon_{s\infty} = \bar{\varepsilon}_{s\infty}\frac{E(7 + 600)}{E(t_0 + \tau_s)}; \quad E(t') = E(28)\sqrt{\frac{t'}{4 + 0.85t'}}$$

$$k'_T = \frac{T}{T_0}\exp\left(\frac{5\,000}{T_0} - \frac{5\,000}{T}\right)$$

$$C_7 = \frac{1}{8}\frac{w}{c}c - 12 \quad \text{and} \quad \text{if } C_7 < 7 \text{ then } C_7 = 7$$

$$\text{if } C_7 > 21 \text{ then } C_7 = 21$$

$$\bar{\varepsilon}_{s\infty} = (1.21 - 0.88y)\,10^{-3}; \quad y = (390z^{-4} + 1)^{-1}$$

$$z = \left[1.25\left(\frac{a}{c}\right)^{1/2} + \frac{1}{2}\left(\frac{g}{s}\right)^2\right]\left(\frac{1 + s/c}{w/c}\right)^{1/3}(f'_c)^{1/2} - 12 \quad \text{if } z \geq 0;$$

$$\text{else } z = 0$$

where f'_c is the cylinder strength after 28 days in ksi ($= 1\,000$ psi $= 6.89$ MN m^{-2}); c is the cement content in kg/m^3; w/c is the water/cement ratio; a/c is the aggregate/cement ratio; g/s is the gravel/sand ratio; s/c is the sand/cement ratio. All ratios are by weight.

k_s is the shape factor with the following values:
1.0 for an infinite plate;
1.15 for an infinite cylinder;
1.25 for an infinite square prism;
1.30 for a sphere;
1.55 for a cube.

T and T_0 are the temperatures in Kelvin, T_0 being the reference temperature $= 23\,°\text{C}$.

The expression for creep is

$$J(t, t') = \frac{1}{E_0} + C_0(t, t') + \bar{C}_d(t, t', t_0) - C_p(t, t', t_0)$$

where $C_0(t, t')$ is the creep without moisture exchange (basic creep);
$\bar{C}_d(t, t', t_0)$ is the creep that develops during moisture exchange (drying creep);
$C_p(t, t', t_0)$ is the creep that occurs after drying has finished.

The above functions are expressed as follows:

$$C_0(t, t') = \frac{\Phi_T}{E_0}(t'_e{}^{-m} + \alpha)(t - t')^{n_T}$$

126

where E_0 represents the asymptotic modulus, and

$$\Phi_T = \Phi_1 C_T = \frac{10^{3n}}{2(28^{-m} + \alpha)} C_T$$

$n = 0.12 + 0.07(1 + 5\,130x^{-6})^{-1}$ for $x > 4$

$n = 0.12$ for $x \leq 4$

$$x = \left[2.1\left(\frac{a}{c}\right)\left(\frac{s}{c}\right)^{-1.4} + 0.1(f_c')^{1.5}\left(\frac{w}{c}\right)^{1.3}\left(\frac{a}{g}\right)^{2.2}\right]a_1$$

$a_1 = 1.0$ for cement types I and II
$a_1 = 0.93$ for cement type III
$a_1 = 1.05$ for cement type IV
$m = 0.28 + f_c'^{-2}$

$$\alpha = \frac{0.025}{w/c}$$

$$t_e' = \int \beta_T(t')\,dt'$$

where β_T expresses the effect of temperature T; if $T = T_0 = 23\,°C$ (reference temperature), then $\beta_T = C_T = 1.0$ and also $t_e' = t'$; $\Phi_T = \Phi_1$ and $n_T = n$.
When the effect of temperature is considered, then

$$\beta_T = \exp\left(\frac{4\,000}{T_0} - \frac{4\,000}{T}\right)$$

$$C_T = c_T \tau_T c_0$$

$$c_T = \frac{19.4}{1 + (100/\hat{T})^{3.5}} - 1; \qquad \tau_T = \frac{1}{1 + 60(t_T')^{-0.69}} + 0.78$$

$$c_0 = \frac{1}{8}\left(\frac{w}{c}\right)^2\left(\frac{a}{c}\right)a_1$$

$$n_T = B_T n; \qquad B_T = \frac{0.25}{1 + (74/\hat{T})^7} + 1; \qquad \hat{T} = T - 253.2$$

$$\bar{C}_d(t, t', t_0) = \frac{\Phi_d'}{E_0} t_e'^{-m/2} k_h \varepsilon_{s\infty} S_d(t, t')$$

$$C_p(t, t', t_0) = c_p k_h'' S_p(t, t_0) C_0(t, t')$$

In these expressions

$$\Phi_d' = \left(1 + \frac{\Delta\tau'}{10}\right)^{-\frac{1}{2}} \Phi_d$$

$\Phi_d = 0.008 + 0.027u$ for $r > 0$
$\Phi_d = 0.008$ for $r \leq 0$

$$u = \frac{1}{1 + 0.7r^{-1.4}}$$

$$r = 56\,000 \left(\frac{s}{a} f_c'\right)^{0.3} \left(\frac{g}{s}\right)^{1.3} \left(\frac{w/c}{\bar{\varepsilon}_{s\infty}}\right)^{1.5} - 0.85$$

$$\Delta\tau' = \int_{t_0}^{t'} \frac{(k_T)^{\frac{5}{4}}}{\tau_s} \, dt$$

$$k_h' \approx 1 - h^{1.5}$$

$$S_d(t, t') = \left(1 + \frac{10\tau_s (k_T')^{\frac{1}{4}}}{t - t'}\right)^{-n'}$$

$$n' = \frac{c_d^n}{K_T^2}; \qquad k_T = 0.42 + 17.6\left[1 + \left(\frac{100}{T}\right)^4\right]^{-1}; c_d = 2.8 - 7.5n$$

$$K_T = 1 + 0.4\left[1 + \left(\frac{93.5}{T}\right)^4\right]^{-1}$$

$$c_p = 0.83; \qquad k_h'' = 1 - h^2$$

$$S_p(t, t_0) = \left(1 + \frac{100}{\Delta\tau}\right)^{-n}; \qquad \Delta\tau = \int_{t_0}^{t'} \frac{dt}{\tau_s}$$

3
Analysis at cross-sectional level

3.1 Theory

The cross-section of a structure is composed of parts with different characteristics (concretes of varying age and creep in reinforced and prestressed structures containing steel); these parts may have even different initial stress values. If the total force and moment quantities loading the cross-section and the initial condition of their action are known, then the aim is to determine the stress variations in the individual parts of the cross-section. The analysis assumes a fully acting cross-section, i.e. prestressed and reinforced-concrete structures in which cracks do not appear.

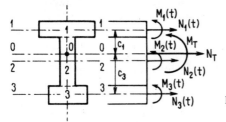

Fig. 3.1 Girder divided into components

Let us consider a cross-section consisting of n parts of different characteristics (the example in Fig. 3.1 shows the case when $n = 3$); this cross-section is subjected to a bending moment M_T and a normal force N_T, acting at point O. In the i-th part of the cross-section, a bending moment $M_i(t)$ variable in time and a normal force $N_i(t)$ are acting. Two conditions of equilibrium apply:

$$\sum_{i=1}^{n} N_i(t) = N_T \tag{3.1}$$

$$\sum_{i=1}^{n} \left[M_i(t) - N_i(t) c_i \right] = M_T \tag{3.2}$$

where c_i is the distance of the centroid of the i-th part from the point O (Fig. 3.1).

129

Since the compatibility of deformation of the individual parts is to be satisfied, the longitudinal strains of the individual components in contact must be equal and maintained during the course of the interaction, and also the curvatures of the deflection lines of all components must remain equal.

The longitudinal strain in time t induced by the stress $\sigma(\tau)$, which is variable in time, is given by the equation (see Eq. (2.25))

$$\varepsilon(t) = \frac{\sigma(t)}{E(t)} - \int_{t_0}^{t} \sigma(\tau) \frac{\partial}{\partial \tau} \left[\frac{1 + \varphi(t, \tau)}{E(\tau)} \right] d\tau \tag{3.3}$$

The stress in the $(i - 1)$-th component in contact with the i-th part is

$$\sigma_{i-1,i}(^{i-1}t) = \frac{N_{i-1}(^{i-1}t)}{A_{i-1}} + \frac{M_{i-1}(^{i-1}t)}{J_{i-1}} c_{i-1,i} \tag{3.4}$$

where $c_{i-1,i}$ is the distance of the centre of the $(i - 1)$-th component from the contact with the i-th part. Similarly, it holds for the stress in the i-th component in contact with the $(i - 1)$-th component:

$$\sigma_{i,i-1}(^{i}t) = \frac{N_i(^{i}t)}{A_i} - \frac{M_i(^{i}t)}{J_i} c_{i,i-1} \tag{3.5}$$

In these equations, A_i designates the cross-sectional areas of the components, and J_i their moments of inertia; ^{i-1}t and ^{i}t designate the age of concrete of the respective component.

The conditions for maintaining equal longitudinal strains at the contact of the adjoining components are:

$$\varepsilon_{i-1,i}(^{i-1}t) - \varepsilon_{i,i-1}(^{i}t) = \Delta\varepsilon_{i-1,i} \tag{3.6}$$

which, considering Eqs. (3.3) to (3.5) and taking into account the effects of shrinkage, can be written in the form

$$\frac{1}{E_{i-1}(^{i-1}t)} \left[\frac{N_{i-1}(^{i-1}t)}{A_{i-1}} + \frac{M_{i-1}(^{i-1}t)}{J_{i-1}} c_{i-1,i} \right] -$$

$$- \int_{^{i-1}t_0}^{^{i-1}t} \left[\frac{N_{i-1}(^{i-1}\tau)}{A_{i-1}} + \frac{M_{i-1}(^{i-1}\tau)}{J_{i-1}} c_{i-1,i} \right]$$

$$\frac{\partial}{\partial^{i-1}\tau} \left[\frac{1 + \varphi_{i-1}(^{i-1}t, {}^{i-1}\tau)}{E(^{i-1}\tau)} \right] d^{i-1}\tau -$$

$$- \varepsilon_{s,i-1} - \frac{1}{E_i(^{i}t)} \left[\frac{N_i(^{i}t)}{A_i} - \frac{M_i(^{i}t)}{J_i} c_{i,i-1} \right] +$$

$$+ \int_{^{i}t_0}^{^{i}t} \left[\frac{N_i(^{i}\tau)}{A_i} - \frac{M_i(^{i}\tau)}{J_i} c_{i,i-1} \right]$$

$$\frac{\partial}{\partial^{i}\tau} \left[\frac{1 + \varphi_i(^{i}t, {}^{i}\tau)}{E(^{i}\tau)} \right] d^{i}\tau + \varepsilon_{s,i} = \Delta\varepsilon_{i-1,i} \tag{3.7}$$

130

where $\Delta\varepsilon_{i-1,i}$ is the difference in strain between both components at the instant when they start to co-operate.

Each part of the cross-section may have, in a general case, its own time coordinate (measured from the instant of casting) and also its own form of the relationship expressing the creep coefficient.

Equations of the type (3.7) are written for each contact and, hence, $(n-1)$ equations are formed.

Assuming the planeness of the cross-section of each component is maintained for small deflections, the curvature of the deflection line is

$$k_i = \frac{\varepsilon_{i,i+1} - \varepsilon_{i,i-1}}{d_i} \tag{3.8}$$

where d_i is the depth of the i-th component $(d_i = c_{i,i-1} + c_{i,i+1})$.

Considering Eqs. (3.3) to (3.5) and (3.8), the condition for equal curvature of the deflection lines of two components is

$$\frac{1}{J_{i-1}}\left\{\frac{M_{i-1}(^{i-1}t)}{E_{i-1}(^{i-1}t)} - \int_{^{i-1}t_0}^{^{i-1}t} M_{i-1}(^{i-1}\tau) \frac{\partial}{\partial^{i-1}\tau}\left[\frac{1 + \varphi_{i-1}(^{i-1}t, {}^{i-1}\tau)}{E(^{i-1}\tau)}\right]d^{i-1}\tau\right\} -$$
$$- \frac{1}{J_i}\left\{\frac{M_i(^it)}{E_i(^it)} - \int_{^it_0}^{^it} M_i(^i\tau) \frac{\partial}{\partial^i\tau}\left[\frac{1 + \varphi_i(^it, {}^i\tau)}{E_i(^i\tau)}\right]d^i\tau\right\} = \Delta k_{i-1,i} \tag{3.9}$$

where $k_{i-1,i}$ is the difference in curvature between both components at the instant when they start to co-operate. Again, the number of independent equations of the type (3.3) will be $(n-1)$. In this way, equations of the type (3.7) together with the conditions of equilibrium (Eqs. (3.1) and (3.2)) represent $2n$ equations for $2n$ unknown functions (normal forces $N_i(t)$ and bending moments $M_i(t)$) in n components.

In case some parts of the cross-section are reinforced or prestressed, the number of unknown quantities is increased by the forces acting in the reinforcement. The equations required for their determination are yielded by the conditions of compatibility of deformation at the points where the reinforcement is placed. The flexural rigidity of the reinforcing bars may be neglected (thus the bending moments in them are zero). Otherwise, the procedure for the analysis of reinforced cross-sections would be unduly toilsome for little or no improvement in the results, considering the accuracy of the creep and shrinkage estimates.

The steel reinforcement in a member subjected to external loading leads to smaller creep deformations than in a member without reinforcement, because the reinforcement restrains creep. Hence, the effect of reinforcement may be expressed by a reduced creep coefficient

$$\varphi' = \frac{1}{\mu n}(1 - e^{-\alpha_0\varphi}) \tag{3.10}$$

131

where (Eq. (3.47) for $c_s = 0$)

$$\alpha_0 = \frac{\mu n}{\mu n + 1}$$

and by an increased modulus $E_c' = E_c(1 + \mu n)$ with the assumption of symmetrical reinforcement.

The derived equations may be solved by means of the methods whose principles are presented in Chapter 2 for the calculation of deformation in a concrete member with known stress history, or for the analysis of the stress-relaxation-type problems (Sects. 2.3 and 2.4). Of these methods, only the method based on the rate-of-creep theory (which, as we know, does not express reality satisfactorily) leads to a closed-form solution.

For a cross-section composed of three parts (Fig. 3.1), when using the rate-of-creep theory, the integral equations of the type (3.7) and (3.9) can be transformed into differential equations of the form

$$\frac{1}{E_1 J_1} [dM_1(t) + M_1(t) \, d\varphi_1(t)] = \frac{1}{E_2 J_2} [dM_2(t) + M_2(t) \, d\varphi_2(t)] \qquad (3.11a)$$

$$\frac{1}{E_2 J_2} [dM_2(t) + M_2(t) \, d\varphi_2(t)] = \frac{1}{E_3 J_3} [dM_3(t) + M_3(t) \, d\varphi_3(t)] \qquad (3.11b)$$

$$\frac{1}{E_1 A_1} [dN_1(t) + N_1(t) \, d\varphi_1(t)] +$$

$$+ \frac{c_{1,2}}{E_1 J_1} [dM_1(t) + M_1(t) \, d\varphi_1(t)] - d\varepsilon_{s1}(t) =$$

$$= \frac{1}{E_2 A_2} [dN_2(t) + N_2(t) \, d\varphi_2(t)] -$$

$$- \frac{c_{2,1}}{E_2 J_2} [dM_2(t) + M_2(t) \, d\varphi_2(t)] - d\varepsilon_{s2}(t) \qquad (3.12a)$$

$$\frac{1}{E_2 A_2} [dN_2(t) + N_2(t) \, d\varphi_2(t)] +$$

$$+ \frac{c_{2,3}}{E_2 J_2} [dM_2(t) + M_2(t) \, d\varphi_2(t)] - d\varepsilon_{s2}(t) =$$

$$= \frac{1}{E_3 A_3} [dN_3(t) + N_3(t) \, d\varphi_3(t)] -$$

$$- \frac{c_{3,2}}{E_3 J_3} [dM_3(t) + M_3(t) \, d\varphi_3(t)] - d\varepsilon_{s3}(t) \qquad (3.12b)$$

Substituting the conditions of equilibrium (Eqs. (3.1) and (3.2)) into these equations, four non-homogeneous differential equations with four unknowns are obtained;

132

they, however, can be solved only by means of iteration, because they contain different creep coefficients $\varphi_1(t)$, $\varphi_2(t)$ and $\varphi_3(t)$. For this reason further simplifications are introduced by using the coefficient of creep affinity $\varkappa$ which is related to the rates of change of the creep coefficients $\varphi_j(t)$ and $\varphi(t)$ according to the relationship

$$\varkappa_j = \frac{d\varphi_j(t)}{dt} : \frac{d\varphi(t)}{dt} \tag{3.13}$$

Here, $\varphi_j(t)$ means the creep coefficient of the component j, where time is reckoned from the instant of its casting, and $\varphi(t)$ is the creep coefficient of the reference component, to which the fundamental time is related.

For example, a member is composed of two components of which the first was cast at a time which is designated as fundamental ($t = 0$), and the second component (j) was cast at time $t = t_j$. The magnitude of the coefficient of affinity after the lapse of another period t_1 is to be determined.

The creep coefficient of the older component for the interval from $t = 0$ until t ($t = t_j + t_1$) is designated $\varphi(t)$, and for the component j

$$\varphi_j(t) = \varphi(t - t_j)$$

If the creep coefficient is calculated according to Eq. (2.8a), then

$$\varphi(t) = \varphi_\infty(1 - e^{-Bt})$$

and also

$$\varphi_j(t) = \varphi_\infty(1 - e^{-B(t-t_j)})$$

Introducing the derivatives of both expressions into Eq. (3.13) we obtain

$$\varkappa_j = \frac{\varphi_j(t)}{\varphi(t)} = e^{Bt_j} = \text{constant}$$

Hence, the ratio of the rates of both creep coefficients does not depend on time t, but solely on time t_j.

The coefficient of affinity, however, is only constant when creep is defined by means of Dischinger's expression: Eq. (2.8a); if it is calculated from Mörsch's expression: Eq. (2.8b), the coefficient of affinity is a function of time and the calculation must use its mean value which is obtained from the values of the coefficients of affinity in the time interval from $t = t_0$ to $t \to \infty$. The coefficients determined in this way for the expression

$$\varphi(t) = \varphi_\infty(1 - e^{-t^{1/2}})^{1/2}$$

are plotted in Fig. 3.2.

For example, at the time when two components, j and k, were made monolithic, the first component is 4 months and the second 8 months old; if the component j is chosen as the reference basis, then $\varkappa_k = 0.75$; if the component k is chosen as the

133

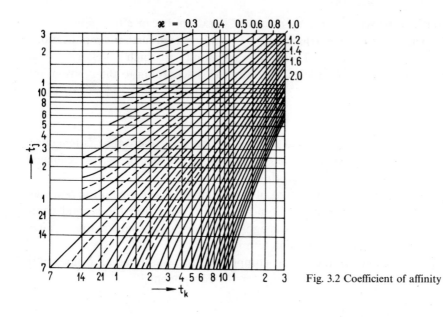

Fig. 3.2 Coefficient of affinity

reference time, then $\varkappa_j = 1.3$. Considering Eq. (3.13) for the designation of Eqs. (3.11) and (3.12), we can write

$$d\varphi_1 = \varkappa_1\, d\varphi; \quad d\varphi_2 = \varkappa_2\, d\varphi; \quad d\varphi_3 = \varkappa_3\, d\varphi \tag{3.14}$$

Let us assume that

$$\varepsilon_{s1}(t) = \frac{\varepsilon_{s1\infty}}{\varphi_{2,1\infty}}\,\varphi_1, \quad \varepsilon_{s2}(t) = \frac{\varepsilon_{s2\infty}}{\varphi_{2,2\infty}}\,\varphi_2, \quad \varepsilon_{s3}(t) = \frac{\varepsilon_{s3\infty}}{\varphi_{2,3\infty}}\,\varphi_3 \tag{3.15}$$

Then, by introducing Eqs. (3.14) and (3.15) into Eqs. (3.11) and (3.12), after rearrangement

$$\frac{1}{E_1 J_1}\left(\frac{\mathrm{d}M_1(t)}{\mathrm{d}\varphi} + M_1(t)\,\varkappa_1\right) = \frac{1}{E_2 J_2}\left(\frac{\mathrm{d}M_2(t)}{\mathrm{d}\varphi} + M_2(t)\,\varkappa_2\right) \tag{3.16a}$$

$$\frac{1}{E_2 J_2}\left(\frac{\mathrm{d}M_2(t)}{\mathrm{d}\varphi} + M_2(t)\,\varkappa_2\right) = \frac{1}{E_3 J_3}\left(\frac{\mathrm{d}M_3(t)}{\mathrm{d}\varphi} + M_3(t)\,\varkappa_3\right) \tag{3.16b}$$

$$\frac{1}{E_1 A_1}\left(\frac{\mathrm{d}N_1(t)}{\mathrm{d}\varphi} + N_1(t)\,\varkappa_1\right) + \frac{c'_1}{E_1 J_1}\left(\frac{\mathrm{d}M_1(t)}{\mathrm{d}\varphi} + M_1(t)\,\varkappa_1\right) -$$

$$-\frac{\varepsilon_{s1\infty}}{\varphi_{2,1\infty}}\,\varkappa_1 = \frac{1}{E_2 A_2}\left(\frac{\mathrm{d}N_2(t)}{\mathrm{d}\varphi} + N_2(t)\,\varkappa_2\right) -$$

$$-\frac{c_2}{E_2 J_2}\left(\frac{\mathrm{d}M_2(t)}{\mathrm{d}\varphi} + M_2(t)\,\varkappa_2\right) - \frac{\varepsilon_{s2\infty}}{\varphi_{2,2\infty}}\,\varkappa_2 \tag{3.17a}$$

$$\frac{1}{E_2 A_2}\left(\frac{dN_2(t)}{d\varphi} + N_2(t)\varkappa_2\right) + \frac{c_2'}{E_2 J_2}\left(\frac{dM_2(t)}{d\varphi} + M_2(t)\varkappa_2\right) -$$

$$-\frac{\varepsilon_{s2\infty}}{\varphi_{2,2\infty}}\varkappa_2 = \frac{1}{E_3 A_3}\left(\frac{dN_3(t)}{d\varphi} + N_3(t)\varkappa_3\right) -$$

$$-\frac{c_3}{E_3 J_3}\left(\frac{dM_3(t)}{d\varphi} + M_3(t)\varkappa_3\right) - \frac{\varepsilon_{s3\infty}}{\varphi_{2,3\infty}}\varkappa_3 \qquad (3.17b)$$

Note: The effect of reinforcement is approximately introduced by reducing the creep coefficient φ to the value φ' in the way indicated earlier. This system can be solved only when there are two equations; however, since only two components of different ages are usually connected into a whole unit in practice, the introduced procedure is applicable.

Other methods of solving the system formed by Eqs. (3.1), (3.2), (3.7) and (3.9) have the character of a numerical solution of integral equations with different rates of approximation. A summary of these methods and their evaluation will be shown on a very simple arrangement: on a cross-section composed of two components with different creep properties and with different initial values of the internal forces; for the sake of simplicity, it is also assumed that both components of the cross-section have the same centroid (for example a two-layer column), and hence, no bending effects arise (Fig. 3.3). The problem is analysed first by the general

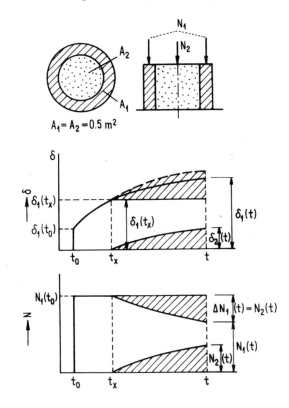

Fig. 3.3 A hollow column

135

Table 3.1

Method	t = 365		t = 730		t = 1825		t = 3650		t = 18250	
	N_2	%	N_2	%	N_2	%	N_2	%	N_2	%
TDM – 120	0.111976	0	0.194611	0	0.268731	0	0.299176	0	0.320476	0
TDM – 100	0.111992	0.0	0.194668	0.0	0.268925	0.1	0.299642	0.2	0.323807	1.0
RCM, Mörsch	0.132877	18.7	0.205463	5.6	0.251830	–6.3	0.266693	–10.9	0.274571	–14.3
EMM	0.101752	–9.1	0.167048	–14.2	0.222237	–17.3	0.244746	–18.2	0.260779	–18.6
AAEMM – $\tilde{R}$	0.104157	–7.0	0.174158	–10.5	0.237814	–11.5	0.267194	–10.7	0.294155	–8.2
AAEMM – S	0.102355	–8.6	0.178368	–8.3	0.249778	–7.1	0.281013	–6.1	0.313339	–2.2
RM 1	0.114888	2.6	0.189987	–2.4	0.253135	–5.8	0.278734	–6.8	0.302801	–5.5
RM 2	0.116278	3.8	0.194731	0.1	0.260062	–3.2	0.282840	–5.5	0.300207	–6.3
ETM $\lambda_1 = 1$	0.112580	0.5	0.194014	–0.3	0.265826	–1.1	0.294933	–1.4	0.315540	–1.5
ETM λ_2^*	0.112019	0.0	0.191889	–1.4	0.260920	–2.9	0.288342	–3.6	0.307429	–4.1

method of time-discretization with varying intervals of time. The results are considered sufficiently accurate to be used as a criterion of the accuracy of other methods [3.1].

Component 1 of the cross-section (external — Fig. 3.3) has been made in time $^1t = 0$ and has been subjected to a normal force $N_1 = 1\,\mathrm{MN}$ in time $^1t_0 = 28$ days. Component 2 (internal) has been cast in time $^1t_x = 180$ days, which is also the origin of time effects for component 2, i.e. $^2t_0 = 0$. The stress is to be determined in both components of the cross-section at time t.

The whole cross-section is subjected to a constant normal force $N = 1\,\mathrm{MN}$. In this case no bending occurs; two unknown functions $N_1(^1t)$ and $N_2(^2t)$ express the time-variation of the normal forces in both components of the cross-section. For the determination of these functions we have the equation of equilibrium (Eq. (3.1)) which, in this case, is

$$N_1(^1t) + N_2(^2t) = N \tag{3.18}$$

and the equations of compatibility of deformations at the contact of both components of the type (Eq. (3.7)); in this case, there is only one such equation of the form

$$\frac{1}{A_1} \left\{ \frac{N_1(^1t)}{E_1(^1t)} - \int_{^1t_0}^{^1t} N_1(^1\tau) \frac{\partial}{\partial^1\tau} \left[\frac{1 + \varphi_1(^1t, {^1\tau})}{E(^1\tau)} \right] d^1\tau \right\} -$$

$$- \frac{1}{A_2} \left\{ \frac{N_2(^2t)}{E_2(^2t)} - \int_{^2t_0}^{^2t} N_2(^2\tau) \frac{\partial}{\partial^2\tau} \left[\frac{1 + \varphi_2(^2t, {^2\tau})}{E_2(^2\tau)} \right] d^2\tau \right\} =$$

$$= \frac{N}{EA_1} [1 + \varphi_1(^1t_x, {^1t_0})] \tag{3.19}$$

The analysis uses the creep coefficient as expressed in accordance with Section 2.6.2. It is assumed that the external part of the cross-section (component 1) hardens in open space ($\varphi_f = 2$, $\varkappa_a = 3$), while the internal component hardens in humid environment ($\varphi_f = 1$, $\varkappa_a = 10$). The effects of temperature, of special type of cement and of a higher content of mixing water ($\varkappa_{tt} = \varkappa_c = \varkappa_b = 1$) are not considered.

The analysis by means of the time-discretization method is based on a substitution of the integrals by summations. The investigated time interval is divided into a series of partial sub-intervals. A step-wise distribution instead of the actual continuous distribution of the desired functions is used. The analysis of the particular case has been carried out with a non-uniform division into 120 sub-intervals, while the same analysis was performed for a division into 100 sub-intervals to estimate the rate of convergency. The results are listed in Table 3.1 (1st and 2nd lines).

Further analyses of the example have been undertaken by means of the following methods:

(a) Analysis based on the rate-of-creep method

According to this theory, the creep coefficient has the following form:

$$\varphi(t, t_0) = \varphi(t) - \varphi(t_0) \tag{3.20}$$

In this case, the problem can be analysed by means of a differential equation. If the creep coefficient satisfies the condition of affinity, there exists an exact solution, whereas in other cases the solution is approximate.

The integral Eq. (3.19) can be transformed into a differential equation (also compare with Eqs. (3.17)). After introducing the creep coefficient (Eq. (3.20)) and assuming $E = $ constant:

$$\frac{1}{E_1 A_1}\left[\frac{dN_1(^1t)}{d^1t} + N_1(^1t)\frac{d\varphi_1(^1t)}{d^1t}\right] -$$

$$- \frac{1}{E_2 A_2}\left[\frac{dN_2(^2t)}{d^2t} + N_2(^2t)\frac{d\varphi_2(^2t)}{d^2t}\right] = 0 \tag{3.21}$$

Following the introduction of the condition of equilibrium (Eq. (3.18)) and of the condition of affinity (which is satisfied only approximately for most of the expressions of the creep coefficient), we obtain one differential equation for the unknown normal force $N_2(t)$ in the younger component of the structure. For the particular case when $A_1 = A_2$, $E_1 = E_2$, we have

$$\frac{dN_2(^1t)}{d^1t} + N_2(^1t)\frac{1 + \varkappa}{2}\frac{d\varphi_1(^1t)}{d^1t} = \frac{N}{2}\frac{d\varphi_1(^1t)}{d^1t} \tag{3.22}$$

137

The general solution after substitution of the initial condition $(^1t = {}^1t_x; N_2(^1t) = 0)$ has the form

$$N_2(^1t) = \frac{N}{1 + \varkappa} \left\{ 1 - e^{-\frac{1+\varkappa}{2}[\varphi_1(^1t) - \varphi_1(^1t_x)]} \right\} \qquad (3.23)$$

Now for this case, the creep coefficient is that of Mörsch:

$$\varphi_1(^1t) = 4(1 - e^{-t\frac{1}{2}})^{\frac{1}{2}} \qquad \text{(compare with Eq. (2.8b))}$$

and the coefficient of affinity has been taken approximately as $\varkappa = 2$. The resulting forces $N_2(^1t)$ are given in Table 3.1, 3rd line.

(b) Effective modulus method

The analysis is analogous to the elastic state, but instead of the current moduli of elasticity, effective moduli are used, viz.

$$E_{\text{eff}_i}(^it, {}^it_0) = \frac{E_i(^it_0)}{1 + \varphi_i(^it, {}^it_0)} \qquad (3.24)$$

The equation of compatibility of deformations at the contact of both components is

$$\frac{N}{E_{\text{eff}1}(^1t, {}^1t_0) A_1} - \frac{N - N_1(^1t)}{E_{\text{eff}1}(^1t, {}^1t_x) A_1} - \frac{N_2(^2t)}{E_{\text{eff}2}(^2t, {}^2t_x) A_2} =$$

$$= \frac{N}{E_{\text{eff}1}(^1t_x, {}^1t_0) A_1} \qquad (3.25)$$

After introducing the condition of equilibrium, the unknown force $N_1(^1t)$ in the older component is eliminated, and we obtain the following formula for the force $N_2(^2t)$ in the younger component (when $A_1 = A_2$):

$$N_2(^2t) = \frac{N}{\dfrac{1}{E_{\text{eff}_1}(^1t, {}^1t_x)} + \dfrac{1}{E_{\text{eff}_2}(^2t, {}^2t_x)}} \frac{1}{E_1(^1t_0)} [\varphi(^1t, {}^1t_0) - \varphi(^1t_x, {}^1t_0)]$$

$$(3.26)$$

The resulting values of $N_2(^1t)$ (assuming $E = $ constant) are listed in Table 3.1, 4th line.

(c) Age-adjusted effective modulus method (Trost-Bažant method)

As stated in Chapter 2, this method is an improvement of the effective modulus method. The equations for the calculation of the internal forces remain unchanged (Eq. (3.26)), but moduli E'' (see Sect. 2.4.3.7) are used instead of moduli E_{eff}. The moduli E'' have been calculated with the aid of an empirical formula for the

138

relaxation function $\tilde{R}_i('t, 't_x)$ [3.2] (Chapter 2) modified as follows (for $E = $ constant)

$$\tilde{R}_i('t, 't_x) =$$

$$= E_i \left\{ \frac{0.992}{1 + \varphi_i('t, 't_x)} - \frac{0.115}{1 + \varphi_i('t, 't - 1)} \left[\frac{1 + \varphi_i('t_x + \xi, 't_x)}{1 + \varphi_i('t, 't - \xi)} - 1 \right] \right\}$$

(3.27)

The age-adjusted effective modulus is given by the relation ([3.2], [3.3])

$$E_i''('t, 't_x) = \frac{E_i - \tilde{R}_i('t, 't_x)}{\varphi_i('t, 't_x)}$$

(3.28)

The resulting forces $N_2('t)$ determined by this procedure are contained in Table 3.1, 5th line.

The age-adjusted effective moduli can be also determined on the basis of the calculation of relaxation with the aid of the function $S(t, t_x, t_0)$ (Section 2.4.3.7). This function expresses the decrease of a unit stress in time t; the stress acts from time t_0 with a constant value until instant t_x and then relaxes until instant t. If the time of the beginning of action of the unit stress and the time of the beginning of relaxation are identical $(t_0 = t_x \Rightarrow S(t, t_x, t_0) = S(t, t_0))$, then the following relationship between the stress function $S(t, t_0)$ and the relaxation function $R(t, t_0)$ applies:

$$S(t, t_0) = 1 - \frac{R(t, t_0)}{E(t_0)}$$

(3.29)

The age-adjusted effective modulus is then calculated from the relation

$$E_i''('t, 't_x, 't_0) = \frac{S_i('t, 't_x, 't_0) E('t_0)}{\varphi_i('t, 't_0) - \varphi_i('t_x, 't_0)}$$

(3.30)

The stress function $S_i('t, 't_x, 't_0)$ is determined by means of the time-discretization method with a division of the interval $\langle 't_x, 't \rangle$ into 100 sub-intervals.

The subsequent procedure is the same as in the foregoing case. The values of the force $N_2('t)$ are given in Table 3.1, 6th line.

(d) Relaxation method

This method is described in detail in Sect. 4.1.5 for the analysis of structures. However, it is also acceptable for an analysis at cross-sectional level.

The investigated time interval is divided into a few sub-intervals and the analysis of each of them consists of two steps.

In the first step, it is assumed that, within the sub-interval, changes of strains are prevented by dummy restraints. Simultaneously, the relaxation of stress takes place in the concrete. In the second step, the cross-section is considered to be elastic and subjected to reactions from dummy restraints of the first step. The resulting stress distribution is approximated by means of the sum of the stress after relaxation and the stress from the second step of the analysis.

The given example is first analysed by means of the relaxation method in its simplest form, when the entire time interval $\langle t_x, t \rangle$ is analysed as a single unit.

It will be recalled that component 1 was subjected to the force N in time t_0, relaxation started at time t_1 and the investigated instant is t. The force N is reduced to $^1\mathfrak{N}_{11}$*) due to relaxation. This means that the force $N - {}^1\mathfrak{N}_{11}$ is taken by the dummy restraint. In the second step of the relaxation method, the elastic structure (in this case both components 1 and 2) is subjected to the loading taken by the dummy restraint, i.e. by the force $N - {}^1\mathfrak{N}_{11}$.

Since $E = E_1 = E_2 = \text{constant}$, and $A_1 = A_2$, both the forces $^1N_{21}$ and $^1N_{22}$, which develop in components 1 and 2 due to the loading by the force $N - {}^1\mathfrak{N}_{11}$, are equal in the second step, viz.

$$^1N_{21} = {}^1N_{22} = \frac{N - {}^1\mathfrak{N}_{11}}{2} \qquad (3.31)$$

The resulting force in component 1 is given by the sum of the forces from the first and second steps.

$$^1N_1 = {}^1\mathfrak{N}_{11} + {}^1N_{21} = {}^1\mathfrak{N}_{11} + \frac{N - {}^1\mathfrak{N}_{11}}{2} \qquad (3.32)$$

At instant t, component 2 is stressed only by the force from the second step

$$^1N_2 = {}^1N_{22} = \frac{N - {}^1\mathfrak{N}_{11}}{2} \qquad (3.33)$$

(Table 3.1, 7th line).

The results obtained from the analysis of the example by means of the relaxation method, with the investigated interval divided into two sub-intervals, are listed in Table 3.1, 8th line.

(e) *Effective time method*

As in the analysis of relaxation (Section 2.4.3.2), the principle of this method is the determination of an instant τ^* (the effective time), from which the interaction forces, maintaining the compatibility of the individual parts of the cross-section in time t, act by their full value. The effective time is determined by employing the rate-of-creep theory and the entire calculation is reduced to the solution of an algebraic equation.

---

*) The upper index 1 indicates a redistribution of forces at the end of the 1st interval. The first subscript designates the 1st or 2nd step of the relaxation method, the second subscript designates the respective part of the cross-section (1 — older component, 2 — younger component).

The analysis consists of three steps: (i) determination of the force N_2, corresponding to the rate-of-creep theory; (ii) calculation of the effective time τ^*; (iii) calculation of N_2 using a creep coefficient of general form.

(i) The force N_2 is determined from the differential equation for the rate-of-creep theory (Eq. (3.22)), which has the solution of Eq. (3.23). The form of creep coefficient may be chosen rather arbitrarily at this stage, because, although affecting τ^*, it does not directly affect the final solution. It might be advisable to use, for example, the third term of the general expression of the creep coefficient in accordance with Sect. 2.6.2 (this term corresponds to the rate-of-creep theory), because of other influencing factors (size and shape of cross-section, humidity of environment, temperature); these factors are not taken into account, for example, in the original expression according to Mörsch. For the given case, the coefficient

$$\varphi_{a1}(^1t) = \lambda \varphi_f \varkappa_h [1 - e^{-(A_d t)\,B_d}] = \lambda \varphi_{ir1}(^1t) \qquad (3.34)$$

has been used where φ_f, $\varkappa_h$, A_d, B_d are the constants for the creep coefficient (Sect. 2.6.2) and λ is a correcting factor (see Eq. (3.40)).

The coefficient of affinity cannot be determined exactly, because Eq. (3.34) does not satisfy the condition of affinity. The coefficient of affinity has been taken approximately as $\varkappa = 2$, by comparing the forms of Eq. (3.34) and of the Mörsch and Dischinger functions.

(ii) When determining the time τ^*, we assume the action of the total loadings of both components of the cross-section (interaction forces N_1, N_2) from instant τ^*. Under this special loading condition, the integral equation (Eq. (3.19)) describing the compatibility of strains at the contact of both components in the investigated time can be modified to an algebraic equation, i.e.

$$\frac{N}{A_1 E_1(^1t_0)}[1 + \varphi_1(^1t,\,^1t_0)] - \frac{N - N_1(^1t)}{A_1 E_1(^1\tau^*)}[1 + \varphi_1(^1t,\,^1\tau^*)] -$$

$$-\frac{N_2(^2t)}{A_2 E_2(^2t_0)}[1 + \varphi_2(^2t,\,^2\tau^*)] = \frac{N}{A_1 E_1(^1t_0)}[1 + \varphi_1(^1t_x,\,^1t_0)] \qquad (3.35)$$

The forces $\tilde{N}_1$ and $\tilde{N}_2$, determined in the analysis by the rate-of-creep theory (Eq. (3.23)), are substituted into the above equation for the forces N_1 and N_2. Also the expression of the coefficient φ_a corresponds to the rate-of-creep theory including the use of the approximate coefficient of affinity. Assuming $E = $ constant, we obtain the following relation:

$$\frac{N}{A_1 E_1}[1 + \varphi_{a1}(^1t) - \varphi_{a1}(^1t_0)] - \frac{N - \tilde{N}_1(^1t)}{A_1 E_1}[1 + \varphi_{a1}(^1t) - \varphi_{a1}(^1\tau^*)] -$$

$$-\frac{\tilde{N}_2(^1t)}{A_2 E_2}[1 + \varkappa\varphi_{a1}(^1t) - \varkappa\varphi_{a1}(^1\tau^*)] =$$

$$= \frac{N}{A_1 E_1}[1 + \varphi_{a1}(^1t_x) - \varphi_{a1}(^1t_0)] \qquad (3.36)$$

which contains only one unknown, namely the creep coefficient $\varphi_{a1}(\tau^*)$. After re-arrangement with $A_1 = A_2$ and $E_1 = E_2$, we have

$$\varphi_{a1}(^1\tau^*) = \frac{\tilde{N}_2(^1t)\,[2 + (1 + \varkappa)\,\varphi_{a1}(^1t)] - N[\varphi_{a1}(^1t) - \varphi_{a1}(^1t_x)]}{\tilde{N}_2(^1t)\,(1 + \varkappa)} \tag{3.37}$$

The instant $^1\tau^*$ is determined by inverting Eq. (3.34):

$$^1\tau^* = \frac{1}{A_d}\left[\ln \frac{1}{1 - \dfrac{\varphi_{a1}(^1\tau^*)}{\varphi_f \varkappa_h \lambda}}\right] \tag{3.38}$$

(iii) In the calculation of the force $N_2(^2t)$ with the application of the general theory of creep, we again assume the loadings of the individual components of the cross-section having constant values. The problem leads to an algebraic equation (Eq. (3.35)), which was used for the determination of time $^1\tau^*$. Now, however, the forces $N_1(^1t)$ and $N_2(^2t)$ are not known. One of the unknowns is eliminated by the condition of equilibrium Eq. (3.18). Following the expression of $N_2(^2t)$ from Eq. (3.35) with $A_1 = A_2$ and $E_1 = E_2$, we have

$$N_2(^2t) = \frac{N[\varphi_1(^1t,\,^1t_0) - \varphi_1(^1t_x,\,^1t_0)]}{2 + \varphi_1(^1t,\,^1\tau^*) + \varphi_2(^2t,\,^2\tau^*)} \tag{3.39}$$

The force $N_2(^1t)$ determined by means of this method without the correcting factor λ ($\lambda = 1$) is given in Table 3.1, 9th line, together with values including the correcting factor

$$\lambda = \frac{\varphi_1(^1t,\,^1t_0)}{\varphi_{ir}(^1t) - \varphi_{ir}(^1t_0)} \tag{3.40}$$

(see Table 3.1, 10th line).

The analysed example demonstrates the practical procedure in employing the individual methods for creep analysis at the cross-sectional level; it also furnishes a comparison of the accuracy of these methods with the "exact solution" obtained on the basis of the time-discretization method.

Referring to Table 3.1, we find that merely using the rate-of-creep theory is evidently unsatisfactory. This theory, however, may be applied with good accuracy in the effective time method which features among the most exact ones.

The effective modulus method may yield rather accurate results in some cases depending on the conditions of application (creep function, length of the investigated time interval).

The age-adjusted effective modulus method approaches the "exact solution" relatively uniformly for all investigated cases. The results depend on the degree to which the assumption of proportionality of the strain variation and the creep coefficient are satisfied. This is related to the fact that the expression of the creep coefficient (Sect. 2.6.2) contains the component of the instantaneous strain which

142

affects the calculation, particularly in the intervals close to the beginning of loading.

The advantage of the relaxation method lies in the possibility of repeated analysis of an elastic structure (the stiffness matrix remains unchanged) and of repeated analysis of relaxation. Hence, this method is good for the preliminary solution of complex problems with the aid of computers. For an application in the design stage, it involves a very easy calculation, particularly if a single time interval can be used.

3.2   Applications

This Section deals with the numerical solutions of some typical creep problems encountered in practical design. The simple methods utilized allow creep effects to be predicted from hand calculations. These calculations facilitate the repeated analyses needed in the preliminary design stage to determine optimum proportions of the structure. The examples, that are presented here, include — besides other approaches — simple analytical solutions based on the rate-of-creep theory for the sake of a qualitative estimation of the importance of the studied phenomena and also as a possible basis for the application of the method of an effective time as presented in Sect. 3.1.

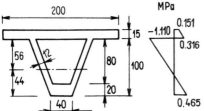

Fig. 3.4 Cross-section of the analysed member

Example

The member whose cross-section is shown in Fig. 3.4 consists of a trough-shaped girder to which the top slab has been cast after five months. It is required to determine the stress which develops due to differential shrinkage of the two components of the girder, three years after their connection. The structure is stored in free air with a relative humidity of 70 %; $E = 36\,000$ MPa $= E_1 = E_2$.

The trough-shaped girder (Component 2):

Its cross-sectional area $A_2 = 0.307\,1$ m^2; moment of inertia $J_2 = 0.027\,74$ m^4; $i_2^2 = 0.090\,3$ m^2; the distance of the bottom edge from the centroid is $c_2' = 0.443\,1$ m

143

and of the top edge $c_2 = 0.556\,9$ m. The effective thickness with consideration of the coefficient $c = 1.5$ (for r.h. $= 70\,\%$) is

$$d_{f2} \doteq \frac{2 \times 0.307\,1 \times 1.5}{4.80} = 0.19 \text{ m}$$

and the final values of the creep and shrinkage coefficients are

$$\varphi_{2,2\infty} = 4.1; \qquad \varepsilon_{s2\infty} = 29 \times 10^{-5}$$

The top slab (Component 1):

The cross-sectional characteristics are:
$A_1 = 0.3 \text{ m}^2$; $J_1 = 0.000\,562\,5 \text{ m}^4$; $c_1' = c_1 = 0.075$ m; $i_1^2 = 0.001\,88 \text{ m}^2$. The effective thickness

$$d_{f1} \doteq \frac{2 \times 2.0 \times 0.15}{4.0}\,1.5 = 0.225 \text{ m}$$

The creep coefficient and the final magnitude of shrinkage are

$$\varphi_{2,1\infty} = 3.9; \qquad \varepsilon_{s1\infty} = 28 \times 10^{-5}$$

As the Mörsch type relation (Eq. (2.8b)) is used for the expression of development of creep and shrinkage, the coefficient of affinity is found in Fig. 3.2. The older component is taken as a reference, and hence the coefficient of affinity of the additionally cast slab will have a value greater than 1.0.

The slab is cast at time $t = 5$ months and the coefficient of affinity $\varkappa_1$ for $t = 5$ months and $t_j = 7$ days is $\varkappa_1 \doteq 2.0$ and $\varkappa_2 = 1.0$, respectively; $c_0 = 0.631\,9$ m. Since only the effect of differential shrinkage of components 1 and 2 is analysed, i.e. $M_0 = N_0 = 0$, the conditions of equilibrium (Eqs. (3.1) and (3.2)) and the strain conditions (Eqs. (3.16) and (3.17)) are

$$N_1(t) + N_2(t) = 0$$

$$M_1(t) + M_2(t) - N_1(t)\,c_0 = 0$$

$$\frac{1}{J_1}\left(\frac{dM_1(t)}{d\varphi} + M_1(t)\,\varkappa_1\right) = \frac{1}{J_2}\left(\frac{dM_2(t)}{d\varphi} + M_2(t)\,\varkappa_2\right)$$

$$\frac{1}{EA_1}\left(\frac{dN_1(t)}{d\varphi} + N_1(t)\,\varkappa_1\right) + \frac{c_1'}{EJ_1}\left(\frac{dM_1(t)}{d\varphi} + M_1(t)\,\varkappa_1\right) -$$

$$- \frac{\varepsilon_{s1\infty}}{\varphi_{2,1\infty}}\,\varkappa_1 = \frac{1}{EA_2}\left(\frac{dN_2(t)}{d\varphi} + N_2(t)\,\varkappa_2\right) -$$

$$- \frac{c_2}{EJ_2}\left(\frac{dM_2(t)}{d\varphi} + M_2(t)\,\varkappa_2\right) - \frac{\varepsilon_{s2\alpha}}{\varphi_{2,2\infty}}\,\varkappa_2$$

144

Eliminating $N_2(t)$ and $M_2(t)$, we obtain

$$\bar{A}_1 \frac{dM_1}{d\varphi} + \bar{A}_2 \frac{dN_1}{d\varphi} + \bar{A}_3 M_1 + \bar{A}_4 N_1 = 0 \tag{3.41a}$$

$$\bar{A}_5 \frac{dM_1}{d\varphi} + \bar{A}_6 \frac{dN_1}{d\varphi} + \bar{A}_7 M_1 + \bar{A}_8 N_1 -$$

$$- E \frac{\varepsilon_{s1\infty}}{\varphi_{2,1\infty}} \varkappa_1 + E \frac{\varepsilon_{s2\infty}}{\varphi_{2,2\infty}} \varkappa_2 = 0 \tag{3.41b}$$

Here

$$\bar{A}_1 = 1 + \frac{J_1}{J_2} \qquad\qquad \bar{A}_2 = -\frac{J_1}{J_2} c_0$$

$$\bar{A}_3 = \varkappa_1 + \varkappa_2 \frac{J_1}{J_2} \qquad\qquad \bar{A}_4 = -\varkappa_2 c_0 \frac{J_1}{J_2}$$

$$\bar{A}_5 = \frac{1}{A_1} \frac{c_1'}{i_1^2} - \frac{1}{A_2} \frac{c_2}{i_2^2} \qquad\qquad \bar{A}_6 = \frac{1}{A_1} + \frac{1}{A_2}\left(1 + \frac{c_2 c_0}{i_2^2}\right)$$

$$\bar{A}_7 = \frac{1}{A_1} \frac{c_1'}{i_1^2} \varkappa_1 - \frac{1}{A_2} \frac{c_2}{i_2^2} \varkappa_2$$

$$\bar{A}_8 = \frac{1}{A_1} \varkappa_1 + \frac{1}{A_2} \varkappa_2 \left(1 + \frac{c_2 c_0}{i_2^2}\right)$$

The constants have the following numerical values:

$$\begin{aligned}
\bar{A}_1 &= 1.020\,3 & \bar{A}_2 &= -0.012\,8 \text{ m} \\
\bar{A}_3 &= 2.020\,3 & \bar{A}_4 &= -0.012\,8 \text{ m} \\
\bar{A}_5 &= 113.255\,5 \text{ m}^{-3} & \bar{A}_6 &= 19.277\,2 \text{ m}^{-2} \\
\bar{A}_7 &= 246.588\,8 \text{ m}^{-3} & \bar{A}_8 &= 22.610\,5 \text{ m}^{-2}
\end{aligned}$$

Substituting into Eqs. (3.41) (the functional symbol (t) is omitted to simplify the formulae), we obtain

$$1.020\,3 \frac{dM_1}{d\varphi} - 0.012\,8 \frac{dN_1}{d\varphi} + 2.020\,3 M_1 - 0.012\,8 N_1 = 0 \tag{a}$$

$$113.255\,5 \frac{dM_1}{d\varphi} + 19.277\,2 \frac{dN_1}{d\varphi} + 246{,}588\,8 M_1 +$$

$$+ 22.610\,5 N_1 - 0.36 \frac{28}{3.9} 2 + 0.36 \frac{29}{4.1} = 0 \tag{a}$$

Putting $M_1 = a_1 e^{\lambda\varphi}$ and $N_1 = a_2 e^{\lambda\varphi}$, the equations become

$$(1.020\,3\lambda + 2.020\,3) a_1 - (0.012\,8\lambda + 0.012\,8) a_2 = 0 \tag{b}$$

145

$$(113.255\ 5\lambda + 246.588\ 8)\ a_1 + (19.277\lambda + 22.610\ 5)\ a_2 -$$

$$- 0.36\frac{28}{3.9}2 + 0.36\frac{29}{4.1} = 0 \tag{b}$$

The characteristic equation of a homogeneous system that is obtained from the condition of zero value of a determinant formed from the coefficients with the unknowns a_1 and a_2, is

$$21.119\ 3\lambda^2 + 66.624\ 8\lambda + 48.838\ 8 = 0$$

The solution yields the roots $\lambda_1 = -1.158\ 4$ and $\lambda_2 = -1.996\ 3$. The unknown constants a_1 and a_2 result from Eqs. (b), viz.

$$\text{for } \lambda_1 = -1.158\ 4 \qquad a_1 = 1; \quad a_2 = -413.112\ 2$$
$$\text{for } \lambda_2 = -1.996\ 3 \qquad a_1 = 1.291\ 8; \quad a_2 = 1.$$

The solution of the homogeneous system resulting from Eqs. (a) may be written in the form

$$M_1 = C_1 a_{1(\lambda_1)}\ e^{\lambda_1\varphi} + C_2 a_{1(\lambda_2)}\ e^{\lambda_2\varphi}$$
$$N_1 = C_1 a_{2(\lambda_1)}\ e^{\lambda_1\varphi} + C_2 a_{2(\lambda_2)}\ e^{\lambda_2\varphi}$$

hence

$$M_1 = C_1\ e^{-1.1584\varphi} + 1.291\ 8C_2\ e^{-1.9963\varphi}$$
$$N_1 = -413.112\ 2C_1\ e^{-1.1584\varphi} + C_2\ e^{-1.9963\varphi}$$

The particular solution is found from Eqs. (a), if $dM_1/d\varphi = dN_1/d\varphi = 0$. Then

$$2.020\ 3M_1 - 0.012\ 8N_1 = 0$$

$$246.588\ 8M_1 + 22.610\ 5N_1 - 0.36\left(\frac{28}{3.9}2 - \frac{29}{4.1}\right) = 0$$

Whence

$$M_1 = 0.000\ 687\ \text{MN m}; \qquad N_1 = 0.108\ 51\ \text{MN}.$$

The total solution is given by the sum of the solutions of the homogeneous equations with particular integral, i.e.

$$M_1 = C_1\ e^{-1.1584\varphi} + 1.291\ 8C_2\ e^{-1.9963\varphi} + 0.000\ 687 \tag{c}$$

$$N_1 = -413.112\ 2C_1\ e^{-1.1584\varphi} + C_2\ e^{-1.9963\varphi} + 0.108\ 51 \tag{c}$$

The constants C_1 and C_2 are evaluated from the condition that, at the connection of the two components of the structure, i.e. at time $t_1 = 5$ months $(= 0.417$ of a year$)$, $M_1 = N_1 = 0$.

146

The creep coefficient for this instant is

$$\varphi(t_1) = 4.1(1 - e^{-0.417\frac{1}{2}})^{\frac{1}{2}} = 2.828$$

and the conditions for the determination of the constants C_1 and C_2 are

$$C_1 e^{-1.1584 \times 2.828} + 1.291\ 8C_2 e^{-1.9963 \times 2.828} + 0.000\ 687 = 0$$

$$-413.112\ 2C_1 e^{-1.1584 \times 2.828} + C_2 e^{-1.9963 \times 2.828} + 0.108\ 51 = 0$$

It follows that $C_1 = 0.006\ 900\ 4$ and $C_2 = -0.207\ 635\ 0$.

At the chosen instant, which in this case is related to the older component, the statical quantities are given by Eqs. (c), into which the already known quantities C_1 and C_2 are substituted.

$$M_1 = 0.006\ 900\ 4\ e^{-1.1584\varphi} - 0.268\ 222\ 9\ e^{-1.9963\varphi} + 0.000\ 687$$

$$N_1 = -2.850\ 639\ 4\ e^{-1.1584\varphi} - 0.207\ 635\ 0\ e^{-1.9963\varphi} + 0.108\ 51$$

After three years, when

$$\varphi = 4.1(1 - e^{-3.0\frac{1}{2}})^{\frac{1}{2}} = 3.72$$

the statical quantities will have the following values:

$$M_1 = 0.000\ 620\ 1\ \text{MN m}; \quad N_1 = 0.070\ 1\ \text{MN}.$$

The bending moment M_2 and the force N_1 are determined from the conditions: Eqs. (3.1) and (3.2); we can write ($N_0 = M_0 = 0$)

$$N_1 = -N_2 = 0.070\ 1\ \text{MN}$$

$$M_2 = -M_1 + N_1 c_0 = 0.043\ 68\ \text{MN m}.$$

The stresses in the outermost fibres of the entire member and at the contacting section of both components are as follows at the top fibre of the member:

$$\sigma_1 = \frac{0.070\ 1}{0.3} - \frac{0.000\ 620\ 1}{0.000\ 562\ 5}0.075 = 0.151\ \text{MPa}$$

at the contacting section:

$$\sigma_1' = \frac{0.070\ 1}{0.3} + \frac{0.000\ 620\ 1}{0.000\ 562\ 5}0.075 = 0.316\ \text{MPa}$$

$$\sigma_2 = \frac{0.070\ 1}{0.307\ 1} - \frac{0.043\ 68}{0.027\ 74} = -1.110\ \text{MPa}$$

at the bottom fibre of the member:

$$\sigma_2' = \frac{0.070\ 1}{0.307\ 1} + \frac{0.043\ 68}{0.027\ 74}0.44 = 0.465\ \text{MPa}$$

The solution has been demonstrated with the example of a composite member which was not subjected to any external load. If the member is loaded, after the connection of the two components and after the hardening of the slab, by an external force N_t acting at the centroid of the component 2 and by a bending moment M_t, both of which are constant with time, and if the two components will have different moduli of elasticity, then Eqs. (3.41) will acquire the form

$$\bar{A}'_1 \frac{dM_1(t)}{d\varphi} + \bar{A}'_2 \frac{dN_1(t)}{d\varphi} + \bar{A}'_3 M_1(t) + \bar{A}'_4 N_1(t) c_0 - \bar{A}'_4 M_t = 0 \qquad (3.42a)$$

$$\bar{A}'_5 \frac{dM_1(t)}{d\varphi} + \bar{A}'_6 \frac{dN_1(t)}{d\varphi} + \bar{A}'_7 M_1(t) + \bar{A}'_8 N_1(t) -$$

$$- n \frac{1}{A_2} \varkappa_2 \left(N_t - \frac{c_2}{i_2^2} M_t \right) - \frac{\varepsilon_{s1\infty}}{\varphi_{2,1\infty}} \varkappa_1 E_1 + \frac{\varepsilon_{s2\infty}}{\varphi_{2,2\infty}} \varkappa_2 E_1 = 0 \qquad (3.42b)$$

The coefficients A'_1 to A'_8, at $n = E_1/E_2$, are

$$\bar{A}'_1 = 1 + n \frac{J_1}{J_2} \qquad\qquad\qquad \bar{A}'_2 = -n \frac{J_1}{J_2} c_0$$

$$\bar{A}'_3 = \varkappa_1 + \varkappa_2 n \frac{J_1}{J_2} \qquad\qquad \bar{A}'_4 = -\varkappa_2 n \frac{J_1}{J_2}$$

$$\bar{A}'_5 = \frac{1}{A_1} \frac{c'_1}{i_1^2} - n \frac{1}{A_2} \frac{c_2}{i_2^2} \qquad\quad \bar{A}'_6 = \frac{1}{A_1} + n \frac{1}{A_2} \left(1 + \frac{c_2 c_0}{i_2^2} \right)$$

$$\bar{A}'_7 = \frac{1}{A_1} \frac{c'_1}{i_1^2} \varkappa_1 - n \frac{1}{A_2} \frac{c_2}{i_2^2} \varkappa_2$$

$$\bar{A}'_8 = \frac{1}{A_1} \varkappa_1 + n \frac{1}{A_2} \varkappa_2 \left(1 + \frac{c_2 c_0}{i_2^2} \right)$$

E_1 and E_2 are the moduli of elasticity of components 1 and 2, respectively.

The analysis is the same as in the preceding case; if, in addition, say, the member in the example is subjected to an external force $N_t = 0.4\,\text{MN}$ and to a bending moment $M_t = 0.12\,\text{MN m}$, the equation is complemented by further absolute terms; if shrinkage is not considered, we have

$$1.020\,3 \frac{dM_1}{d\varphi} - 0.012\,8 \frac{dN_1}{d\varphi} + 2.020\,3 M_1 - 0.012\,8 N_1 +$$

$$+ 0.012\,8 \times 0.12 = 0 \qquad\qquad (a)$$

$$113.255\,5 \frac{dM_1}{d\varphi} + 19.277\,2 \frac{dN_1}{d\varphi} + 246.588\,8 M_1 + 22.610\,5 N_1 -$$

$$- 3.256\,6(0.4 - 0.616\,5 \times 0.12) = 0 \qquad\qquad (b)$$

Here, $1/A_2 = 3.256\,6$ and $c_2/i_2^2 = 0.616\,5$.

148

Since the homogeneous equations are the same as in the case when no external forces acted, their solution is known and it suffices to determine the particular integral from the equations

$$2.020\ 3M_1 - 0.012\ 8N_1 + 0.001\ 537\ 2 = 0$$

$$246.588\ 8M_1 + 22.610\ 5N_1 - 1.061\ 719\ 2 = 0$$

The solution yields the values $N_1 = 0.041$ MN; $M_1 = -0.000\ 5$ MN m. Further analysis will be the same as in the foregoing example; when determining the constants C_1 and C_2, it should be observed that, after hardening and imposition of the loadings M_t and N_t (for example, 6 months measured according to the reference time, i.e. from the second component), M_1 must equal M_t and N_1 must equal N_t.

The same procedure is adopted, if at the time of connection of both components, one is already subjected to a force N_0, or to a bending moment M_0.

Modification of the analysis by means of the relaxation coefficient

To analyse the redistribution of stresses, the relaxation-coefficient method may be employed. This method is convenient primarily, because the solution of a set of differential equations does not arise and only a set of algebraic equations is dealt with; secondly, the reinforcement of the individual components may be considered simply. The procedure in the deduction of the fundamental equations is similar to that used in the foregoing Section, since it is inherent in the expression for the strain conditions. In the relaxation theory (Sect. 2.4.3.5), the strain at any instant may be expressed by the equation

$$\varepsilon(t) = \frac{\sigma_0}{E}(1 + \varphi(t, t_0)) + \frac{\sigma(t) - \sigma_0}{E}(1 + \bar{\varrho}\varphi(t, t_0)) + \varepsilon_s(t) =$$

$$= \frac{\sigma_0}{E}(1 - \bar{\varrho})\,\varphi(t, t_0) + \frac{\sigma(t)}{E}(1 + \bar{\varrho}\varphi(t, t_0)) + \varepsilon_s(t) \tag{3.43}$$

Re-considering the member portrayed in Fig. 3.1, conditions of equilibrium and the deformation conditions can be written in the form

$$N_1(t) + N_2(t) + N_3(t) = N_t \tag{3.44a}$$

$$M_1(t) + M_2(t) + M_3(t) - N_1(t)\,c_0 + N_3(t)\,c_0' = M_t \tag{3.44b}$$

considering N_t at the centroid of the component 2.

$$\frac{1}{E_1 J_1}\left[M_{01}(1 - \bar{\varrho}_1)\,\varphi_1(t, t_0) + M_1(t)\,(1 + \bar{\varrho}_1\varphi_1(t, t_0))\right] =$$

$$= \frac{1}{E_2 J_2}\left[M_{02}(1 - \bar{\varrho}_2)\,\varphi_2(t, t_0) + M_2(t)\,(1 + \bar{\varrho}_2\varphi_2(t, t_0))\right] \tag{3.45a}$$

$$\frac{1}{E_2 J_2} \left[M_{02}(1 - \bar{\varrho}_2) \varphi_2(t, t_0) + M_2(t)(1 + \bar{\varrho}_2 \varphi_2(t, t_0)) \right] =$$

$$= \frac{1}{E_3 J_3} \left[M_{03}(1 - \bar{\varrho}_3) \varphi_3(t, t_0) + M_3(t)(1 + \bar{\varrho}_3 \varphi_3(t, t_0)) \right] \qquad (3.45b)$$

$$\frac{1}{E_1 A_1} \left[N_{01}(1 - \bar{\varrho}_1) \varphi_1(t, t_0) + N_1(t)(1 + \bar{\varrho}_1 \varphi_1(t, t_0)) \right] +$$

$$+ \frac{c_1'}{E_1 J_1} \left[M_{01}(1 - \bar{\varrho}_1) \varphi_1(t, t_0) + M_1(t)(1 + \bar{\varrho}_1 \varphi_1(t, t_0)) \right] - \varepsilon_{s1}(t) =$$

$$= \frac{1}{E_2 A_2} \left[N_{02}(1 - \bar{\varrho}_2) \varphi_2(t, t_0) + N_2(t)(1 + \bar{\varrho}_2 \varphi_2(t, t_0)) \right] -$$

$$- \frac{c_2}{E_2 J_2} \left[M_{02}(1 - \bar{\varrho}_2) \varphi_2(t, t_0) + M_2(t)(1 + \bar{\varrho}_2 \varphi_2(t, t_0)) \right] - \varepsilon_{s2}(t) \qquad (3.46a)$$

$$\frac{1}{E_2 A_2} \left[N_{02}(1 - \bar{\varrho}_2) \varphi_2(t, t_0) + N_2(t)(1 + \bar{\varrho}_2 \varphi_2(t, t_0)) \right] +$$

$$+ \frac{c_2'}{E_2 J_2} \left[M_{02}(1 - \bar{\varrho}_2) \varphi_2(t, t_0) + M_2(t)(1 + \bar{\varrho}_2 \varphi_2(t, t_0)) \right] - \varepsilon_{s2}(t) =$$

$$= \frac{1}{E_3 A_3} \left[N_{03}(1 - \bar{\varrho}_3) \varphi_3(t, t_0) + N_3(t)(1 + \bar{\varrho}_3 \varphi_3(t, t_0)) \right] -$$

$$- \frac{c_3}{E_3 J_3} \left[M_{03}(1 - \bar{\varrho}_3) \varphi_3(t, t_0) + M_3(t)(1 + \bar{\varrho}_3 \varphi_3(t, t_0)) \right] - \varepsilon_{s3}(t) \qquad (3.46b)$$

In these equations, M_{01}; M_{02}; M_{03} and N_{01}; N_{02}; N_{03} designate the static values which act on the individual components at the time of connection (usually prestressing and dead-weight).

The relaxation coefficient $\bar{\varrho}$ is determined from Fig. 2.23. If it is also desired to consider the effect of the reinforcements of the individual components which influence creep, the following coefficient [3.4, 3.5, 3.6]

$$\alpha_0 = \frac{\delta_c}{\delta_c + \delta_s}$$

is introduced, by which the quantities φ_∞ are multiplied, so that $\bar{\varphi}'_\infty = \alpha_0 \bar{\varphi}_\infty$ is necessary in the determination of the relaxation coefficient; these values are substituted for the values $\bar{\varphi}_\infty$ in Fig. 2.23. In the formula for α_0, δ_c is the strain of concrete induced by a unit force acting in the steel, and δ_s is the strain of steel induced by the same unit force. If the reinforcement is situated eccentrically in the cross-section:

$$\delta_c = \frac{i^2 + c_s^2}{A_c E_c i^2}; \qquad \delta_s = \frac{1}{E_s A_s}$$

and, for asymmetrically arranged reinforcement, we have

$$\alpha_0 = \frac{\dfrac{i^2 + c_s^2}{i^2}\, n\mu}{\dfrac{i^2 + c_s^2}{i^2}\, n\mu + 1} \tag{3.47}$$

Here i is the radius of inertia of the concrete section (without reinforcement);

c_s is the distance of the reinforcement from the centroid of the concrete cross-section;

E_c and E_s are the moduli of elasticity of concrete and steel, respectively;

A_c and A_s are the cross-sectional areas of concrete (without reinforcement) and of steel, respectively.

For symmetrical reinforcement, $\delta_c = 1/E_c A_c$, hence

$$\alpha_0 = \frac{\mu n}{1 + \mu n}$$

where $\mu = A_s/A_c$; $n = E_s/E_c$

Usually, μ is 0.006 to 0.03, which, at $n = 10$, yields $\alpha_0 \doteq 0.06$ to 0.23; consequently, for the highest $\bar{\varphi}_\infty = 3.5$, i.e. as given in Fig. 2.23, $\bar{\varphi}_\infty = (0.06 \div 0.23)\,3.5 \doteq$ $\doteq 0.21 \div 0.81$, hence < 1.0; in this case, $(\alpha_0 \bar{\varphi}_\infty = 1.0)$, $\bar{\varrho} \doteq 0.81$ for $\tau_0 = 28$ days; at low reinforcement, $\alpha_0 \bar{\varphi}_\infty$ may be considered equal to zero. If two components of different ages are connected, Eqs. (3.44) to (3.46) are modified into the form

$$N_1(t) + N_2(t) = N_t \tag{3.48a}$$

$$M_1(t) + M_2(t) - N_1 c_0 = M_t \tag{3.48b}$$

$$\left[(1 + \bar{\varrho}_1 \varphi_1(t, t_0)) + \frac{E_1 J_1}{E_2 J_2}(1 + \bar{\varrho}_2 \varphi_2(t, \tau_0))\right] M_1 -$$

$$- \frac{E_1 J_1}{E_2 J_2}(1 + \bar{\varrho}_2 \varphi_2(t, t_0))\, c_0 N_1 + (1 - \bar{\varrho}_1)\,\varphi_1(t, t_0)\, M_{01} -$$

$$- \frac{E_1 J_1}{E_2 J_2}(1 - \bar{\varrho}_2)\,\varphi_2(t, t_0)\, M_{02} - \frac{E_1 J_1}{E_2 J_2}(1 + \bar{\varrho}_2 \varphi_2(t, t_0))\, M_t = 0 \tag{3.49}$$

$$\left[\frac{c_1'}{A_1 i_1^2}(1 + \bar{\varrho}_1 \varphi_1(t, t_0)) - \frac{E_1 c_2}{E_2 A_2 i_2^2}(1 + \bar{\varrho}_2 \varphi_2(t, t_0))\right] M_1 +$$

$$+ \left[\frac{1}{A_1}(1 + \bar{\varrho}_1 \varphi_1(t, t_0)) + \frac{E_1}{E_2}\left(\frac{1}{A_2} + \frac{c_2 c_0}{A_2 i_2^2}\right)(1 + \bar{\varrho}_2 \varphi_2(t, t_0))\right] N_1 +$$

$$+ \frac{1}{A_1}(1 - \bar{\varrho}_1)\,\varphi_1(t, t_0)\, N_{01} - \frac{E_1}{E_2 A_2}(1 - \bar{\varrho}_2)\,\varphi_2(t, t_0)\, N_{02} +$$

$$+ \frac{c_1'}{A_1 i_1^2}(1 - \bar{\varrho}_1)\,\varphi_1(t, t_0)\, M_{01} + \frac{E_1 c_2}{E_2 A_2 i_2^2}(1 - \bar{\varrho}_2)\,\varphi_2(t, t_0)\, M_{02} -$$

$$-\frac{E_1}{E_2 A_2}(1 + \bar{\varrho}_2 \varphi_2(t, t_0)) N_t + \frac{E_1 c_2}{E_2 A_2 i_2^2}(1 + \bar{\varrho}_2 \varphi_2(t, t_0)) M_t -$$
$$- E_1 \varepsilon_{s1} + E_1 \varepsilon_{s2} = 0 \qquad (3.50)$$

Example

The member portrayed in the preceding example is made by casting a slab on a trough-shaped girder 6 months old. The forces are to be determined, which develop in the individual components due to shrinkage only, 2.5 years after the connection of both components. The structure is stored in free air with r.h. = 70%; the elasticity moduli of both components E_c = 36 000 MPa. Creep and shrinkage are determined according to the CEB–FIP Recommendations (Sect. 2.6),

$$M_t = M_{01} = M_{02} = N_t = N_{01} = N_{02} = 0$$

Trough–shaped girder (Component 2):

$$A_2 = 0.307\,1 \text{ m}^2, \quad J_2 = 0.027\,74 \text{ m}^4, \quad i_2^2 = 0.090\,33 \text{ m}^3$$
$$c_2 = 0.556\,9 \text{ m}, \quad c_0 = 0.631\,9 \text{ m}$$
$$d_{f2} = \frac{2 \times 0.307\,1}{4.80} = 0.13 \qquad \alpha_{02} = 0.2$$
$$k_c = 2.3, \quad k_b = 1.0, \quad k_e \doteq 0.95, \quad k_d = 0.6$$

k_t (considered since the connection, i.e., for 2.5 years) = 0.85

$$\bar{\varphi}_\infty = 0.95 \times 1.0 \times 2.3 = 2.19; \quad \bar{\varphi}'_\infty = 0.2 \times 2.19 \doteq 0.44$$

For $\bar{\varphi}'_\infty = 0.44$ and $\tau_0 = 0.5$ of a year = 180 days, $\bar{\varrho} \doteq 0.88$ (Fig. 2.23)

$$\varphi_2(t, t_0) = \varphi_\infty k_d k_t = 2.19 \times 0.6 \times 0.85 = 1.12$$
$$\varepsilon_{s0} = 27.5 \times 10^{-5}, \quad \mu n = 25, \quad k_b = 1.0, \quad k'_e \doteq 0.95, \quad k_p \doteq 0.8$$

k_t (considered 2.5 years after connection), i.e.

$$k_t = k_{t(3.0 \text{ years})} - k_{t(0.5 \text{ year})} = 0.92 - 0.70 = 0.22$$
$$\varepsilon_{s2} = 27.5 \times 1.0 \times 0.95 \times 0.8 \times 0.22 \times 10^{-5} = 4.6 \times 10^{-5}$$

Cast-in-situ slab (Component 1)

$$A_1 = 0.3 \text{ m}^2, \quad J_1 = 0.000\,562\,5 \text{ m}^4, \quad i_1^2 = 0.001\,88 \text{ m}^2,$$
$$c'_1 = 0{,}075 \text{ m}, \quad d_{f1} = \frac{2 \times 2.0 \times 0.15}{4.0} = 0.15, \quad \alpha_1 = 0.1$$
$$k_c = 2.3, \quad k_b = 1.0, \quad k_e \doteq 1.90, \quad k_d = 1.8$$

$k_t = 0.85$ (considered since the connection)

$$\bar{\varphi}_\infty = 0.90 \times 1.0 \times 2.3 = 2.07, \quad \bar{\varphi}'_\chi = 0.1 \times 2.07 \doteq 0.2$$

For $\bar{\varphi}'_\infty = 0.2$ and $\tau_0 = 1$ day, $\bar{\varrho} \doteq 0.58$

$$\varphi_1(t, t_0) = 2.07 \times 1.8 \times 0.85 = 3.17$$

$$\varepsilon_{s0} = 27.5 \times 10^{-5}, \quad \mu n = 10, \quad k_b = 1.0, \quad k'_e \doteq 0.90, \quad k_p \doteq 0.9$$

k_t (is considered 2.5 years after connection), i.e.

$$k_t = k_{t(2.5\ years)} = 0.90$$

$$\varepsilon_{s1} = 27.5 \times 1.0 \times 0.90 \times 0.90 \times 0.90 \times 10^{-5} = 20.5 \times 10^{-5}$$

Substituting into Eqs. (3.49) and (3.50) and regarding the conditions of equilibrium: Eqs. (3.48a) and (3.48b), then re-arranging and expressing numerically

$$2.880\,0\,M_1 - 0.025\,5\,N_1 = 0$$

$$338.716\,3\,M_1 - 41.191\,2\,N_1 - 5.724 = 0$$

for which the solution is

$$N_1 = -0.129 \text{ MN} \quad \text{and} \quad M_1 = 0.001\,15 \text{ MN m}$$

The conditions of equilibrium yield

$$N_2 = -0.129 \text{ MN}; \quad M_2 = 0.081 \text{ MN m}$$

Reinforced cross-sections

In reinforced cross-sections, the stress is transferred from the concrete to the reinforcing steel bars due to the external long-term loading and due to shrinkage so that the stress in the reinforcing steel increases. On the other hand, the stress in the concrete decreases, which may lead to the development of tensile stresses when the external compressive forces are small.

The redistribution of stresses may be analysed by means of the methods given in the foregoing Sections, and also the deduced equations may be used. The analysis will be demonstrated on the example of a rectangular section asymmetrically reinforced by two rows of bars (Fig. 3.5); first, by means of the rate-of-creep method and, secondly, by means of its modified form utilizing the mean stress value.

(a) *Analysis by means of the rate-of-creep theory*

The cross-section is divided into three components, of which the first is represented by the concrete section, the second by the top reinforcement and the third by the bottom reinforcement. The components are subjected to forces $N_c(t)$

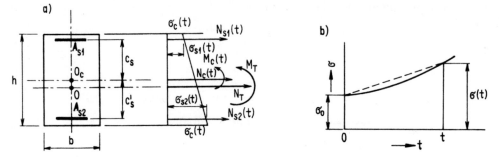

Fig. 3.5 Creep and shrinkage of a reinforced cross-section
(a) designation of quantities; (b) simplified solution by means of mean value

and $M_c(t)$, $N_{s1}(t)$ and $N_{s2}(t)$. $M_{s2}(t)$ and $M_{s1}(t)$ are neglected, because the flexural rigidity of steel is small. Then the conditions of equilibrium are (with N_t acting at the point 0_c)

$$N_c(t) + N_{s1}(t) + N_{s2}(t) = N_t \tag{3.51a}$$

$$M_c(t) - N_{s1}(t) c_s + N_{s2}(t) c'_s = M_t \tag{3.51b}$$

and the conditions of deformation together with consideration of shrinkage of concrete, yield

$$\frac{1}{E_s A_{s1}} \frac{dN_{s1}(t)}{d\varphi} = \frac{1}{E_c A_c} \left(\frac{dN_c(t)}{d\varphi} + N_c(t) \right) -$$

$$- \frac{c_s}{E_c J_c} \left(\frac{dM_c(t)}{d\varphi} + M_c(t) \right) - \frac{\varepsilon_{s\infty}}{\varphi_{2\infty}} \tag{3.52a}$$

$$\frac{1}{E_s A_{s2}} \frac{dN_{s2}(t)}{d\varphi} = \frac{1}{E_c A_c} \left(\frac{dN_c(t)}{d\varphi} + N_c(t) \right) +$$

$$+ \frac{c'_s}{E_c J_c} \left(\frac{dM_c(t)}{d\varphi} + M_c(t) \right) - \frac{\varepsilon_{s\infty}}{\varphi_{2\infty}} \tag{3.52b}$$

After re-arrangement, we can write

$$\bar{A}''_1 \frac{dN_{s1}}{d\varphi} + \bar{A}''_2 \left(\frac{dN_{s2}}{d\varphi} + N_{s2} \right) + \bar{A}''_3 N_{s1} - \mu_1 n \left(N_t + \frac{c_s}{i^2} M_t \right) +$$

$$+ E_s A_{s1} \frac{\varepsilon_{s\infty}}{\varphi_{2\infty}} = 0 \tag{3.53a}$$

$$\bar{A}''_4 \left(\frac{dN_{s1}}{d\varphi} + N_{s1} \right) + \bar{A}''_5 \frac{dN_{s2}}{d\varphi} + \bar{A}''_6 N_{s2} - \mu_1 n \left(N_t + \frac{c'_s}{i^2} M_t \right) +$$

$$+ E_s A_{s2} \frac{\varepsilon_{s\infty}}{\varphi_{2\infty}} = 0 \tag{3.53b}$$

154

The constants have the values

$$\bar{A}_1'' = 1 + \mu_1 n \left(1 + \frac{c_s^2}{i^2} \right) \qquad \bar{A}_2'' = \mu_1 n \left(1 - \frac{c_s c_s'}{i^2} \right)$$

$$\bar{A}_3'' = \mu_1 n \left(1 + \frac{c_s^2}{i^2} \right) \qquad \bar{A}_4'' = \mu_2 n \left(1 - \frac{c_s c_s'}{i^2} \right)$$

$$\bar{A}_5'' = 1 + \mu_2 n \left(1 + \frac{c_s'^2}{i^2} \right) \qquad \bar{A}_6'' = \mu_2 n \left(1 + \frac{c_s'^2}{i^2} \right)$$

where $\mu_1 = A_{s1}/A_c$; $\mu_2 = A_{s2}/A_c$

The analysis will be demonstrated on a rectangular cross-section, which is simply reinforced by bars of cross-sectional area A_{s1} and which is subjected to a force $N_t = -P$, acting in the axis of the reinforcement; it is assumed that the centroid O of the entire section differs insignificantly from the centroid O_c of the concrete section: then, $M_t = -N_t c_s$.

The force N_{s1} in the reinforcement is obtained from the solution of one of the two equations; the first equation yields

$$\bar{A}_1'' \frac{dN_{s1}}{d\varphi} + \bar{A}_3'' N_{s1} - \mu_1 n \left(1 + \frac{c_s^2}{i^2} \right) N_t = 0$$

The homogeneous equation yields

$$N_{s1,1} = C\, e^{-\frac{\bar{A}_3''}{\bar{A}_1''}\varphi}$$

The particular solution is $N_{s1,0} = N_t$, because $\mu_1 n \left(1 + \frac{c_s^2}{i^2} \right) = \bar{A}_3''$. The total solut-

ion is $N_{s1} = C\, e^{-\frac{\bar{A}_3''}{\bar{A}_1''}\varphi} + N_t$ and since $N_{s1} = 0$ for $\varphi = 0$, $C = -N_t$. Then

$$N_{s1} = N_t(1 - e^{-\frac{\bar{A}_3''}{\bar{A}_1''}\varphi}) = -P(1 - e^{-\frac{\bar{A}_3''}{\bar{A}_1''}\varphi})$$

and the condition of equilibrium of forces yields

$$N_c = N_t - N_{s1} = N_t\, e^{-\frac{\bar{A}_3''}{\bar{A}_1''}\varphi} = -P\, e^{-\frac{\bar{A}_3''}{\bar{A}_1''}\varphi}$$

The root of the linear characteristic equation is

$$\frac{\bar{A}_3''}{\bar{A}_1''} = \frac{\mu n \left(1 + \frac{c_s^2}{i^2} \right)}{1 + \mu n \left(1 + \frac{c_s^2}{i^2} \right)} = \frac{\mu n \frac{i^2 + c_s^2}{i^2}}{\frac{i^2 + c_s^2}{i^2}\mu n + 1} = \alpha_0$$

This is the coefficient corresponding to Eq. (3.47); it expresses the effect of reinforcement on creep, because the value of the creep coefficient changes in accordance with the resulting expression for N_c.

If the reinforcement of the cross-section is symmetrical, then the reduced creep coefficient φ' may be determined by comparing the deformations due to creep in an unreinforced cross-section and that of a reinforced section.

If a cross-section of plain concrete is subjected to a force $P = -N_c$, then the strain due to creep is

$$\varepsilon_c = \frac{P}{E_c A_c} \varphi$$

and for a symmetrically reinforced section it is

$$\varepsilon_c = \frac{1}{E_c A_c} \int_0^{\varphi} \left(\frac{dN_c}{d\varphi} + N_c \right) d\varphi =$$

$$= -\frac{P}{E_c A_c} \int_0^{\varphi} \left(\frac{d}{d\varphi} e^{-\alpha_0 \varphi} + e^{-\alpha_0 \varphi} \right) d\varphi =$$

$$= -\frac{P}{E_c A_c} \frac{1}{\mu n} \left(1 - e^{-\alpha_0 \varphi} \right) = -\frac{P}{E_c A_c} \varphi'$$

Hence, it suffices to introduce the effect of the reinforcement by means of the reduced creep coefficient φ' and to analyse the member as if it were made of plain concrete. Since, however, the strain characteristics of a reinforced cross-section are also different, the effective modulus $E_c' = E_c(1 + n\mu)$ should be introduced instead of E_c.

(b) *Simplification of the analysis by use of the mean stress value* (see Sect. 2.4.3.8)

The relationship between strain and stress is expressed in the rate-of-creep theory by means of Eq. (2.30)

$$\varepsilon(t) = \frac{\sigma(t)}{E_c} + \frac{1}{E_c} \int_{\tau_1}^{t} \sigma(\tau) \frac{d\varphi(\tau)}{d\varphi} d\tau$$

Using the mean stress value theorem, we can write

$$\varepsilon(t) = \frac{\sigma(t)}{E_c} + \frac{1}{E_c} \varphi(t) \overset{t}{\underset{\tau_1}{S}}(\sigma)$$

where $\overset{t}{\underset{\tau_1}{S}}(\sigma)$ is the mean stress value in the interval $\langle \tau_1, t \rangle$.

Comparative calculations indicate that the relationship between the stress $\sigma(t)$ in concrete and the creep coefficient $\varphi(t)$ is nearly linear; it follows from [3.7] that, if $\varphi_t \leq 2.0$ and the steel ratio $\mu \leq 0.03$, the difference between the stress calculated "accurately" and the calculation using the mean value does not exceed 5 to 6 per cent.

Assuming a linear relationship between $\sigma(t)$ and $\varphi(t)$, we can write

$$\varepsilon(t) = \frac{\sigma(t)}{E_c} + \frac{\sigma_0 + \sigma(t)}{2E_c} \varphi(t) \tag{3.54}$$

where σ_0 is substituted for $\sigma(\tau_1)$, designating the initial stress (Fig. 3.5b). Using the above expression for the calculation of strain, the conditions of compatibility (Eqs. (3.52a) and (3.52b)) give

$$\frac{1}{E_s A_{s1}} N_{s1} = \frac{1}{E_c A_c}\left(N_c + \frac{N_0 + N_c}{2}\varphi\right) -$$

$$- \frac{c_s}{E_c J_c}\left(M_c + \frac{M_0 + M_c}{2}\varphi\right) - \varepsilon_s \qquad (3.55a)$$

$$\frac{1}{E_s A_{s2}} N_{s2} = \frac{1}{E_c A_c}\left(N_c + \frac{N_0 + N_c}{2}\varphi\right) +$$

$$+ \frac{c_s'}{E_c J_c}\left(M_c + \frac{M_0 + M_c}{2}\varphi\right) - \varepsilon_s \qquad (3.55b)$$

which, after substitution from the conditions of equilibrium, yield

$$\bar{A}_1''' N_{s1} + \bar{A}_2''' N_{s2} - \mu_1 n\left(1 + \frac{\varphi}{2}\right)\left(N_t - \frac{c_s}{i^2} M_t\right) -$$

$$- \mu_1 n \frac{\varphi}{2}\left(N_0 - \frac{c_s}{i^2} M_0\right) + E_s A_{s1}\varepsilon_s = 0 \qquad (3.56a)$$

$$\bar{A}_3''' N_{s1} + \bar{A}_4''' N_{s2} - \mu_2 n\left(1 + \frac{\varphi}{2}\right)\left(N_t + \frac{c_s'}{i^2} M_t\right) -$$

$$- \mu_2 n \frac{\varphi}{2}\left(N_0 + \frac{c_s'}{i^2} M_0\right) + E_s A_{s2}\varepsilon_s = 0 \qquad (3.56b)$$

Here

$$\bar{A}_1''' = 1 + \mu_1 n\left(1 + \frac{\varphi}{2}\right)\left(1 + \frac{c_s^2}{i^2}\right)$$

$$\bar{A}_2''' = \mu_1 n\left(1 + \frac{\varphi}{2}\right)\left(1 - \frac{c_s c_s'}{i^2}\right)$$

$$\bar{A}_3''' = \mu_2 n\left(1 + \frac{\varphi}{2}\right)\left(1 - \frac{c_s c_s'}{i^2}\right)$$

$$\bar{A}_4''' = 1 + \mu_2 n\left(1 + \frac{\varphi}{2}\right)\left(1 + \frac{c_s'^2}{i^2}\right)$$

Example

The cross-section portrayed in Fig. 3.6 is symmetrically reinforced by bars of a cross-sectional area $A_s = 99.14 \text{ cm}^2$ and it is subjected to a compressive force $N_t = -3.8 \text{ MN}$ acting at the centre of the cross-section; the force is applied to the section 28 days after casting.

157

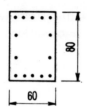

Fig. 3.6 Cross-section of the analysed
reinforced concrete member

$E_s = 0.21 \times 10^6$ MPa; $n = 5.8$

$A_c = 0.48$ m^2; $\mu \doteq 0.02$

$\varepsilon_{s\infty} = 28 \times 10^{-5}$; $\varphi_{2\infty} = 3.9$

The change of forces in the reinforcement, which takes place from $\tau_1 = 28$ days until $t = \infty$, is to be determined.

The creep coefficient for the period $\langle 28 \text{ days}, \infty \rangle$ is

$$\varphi = 3.9\left[1 - (1 - e^{-0.0767^{\frac{1}{2}}})^{\frac{1}{4}}\right] = 1.98$$

and hence creep is

$$\varepsilon_s = \frac{28}{3.9} 1.98 \times 10^{-5} = 14.22 \times 10^{-5}$$

More accurate analysis

With reference to Eq. (3.53a), for $N_{s2} = 0$ and $c_s = 0$

$$\bar{A}_1'' \frac{dN_s}{d\varphi} + \bar{A}_3'' N_s - \mu n N_t + E_s A_s \frac{\varepsilon_{s\infty}}{\varphi_\infty} = 0$$

$$\bar{A}_1'' = 1 + 0.02 \times 5.8 = 1.116, \quad \bar{A}_2'' = 0.02 \times 5.8 = 0.116$$

$$-\mu n N_t + E_s A_s \frac{\varepsilon_{s\infty}}{\varphi_\infty} = 0.02 \times 5.8 \times 3.8 +$$

$$+ 0.21 \times 10^6 \times 0.009\,914 \frac{28}{3.9} 10^{-5} = 0.590\,27 \text{ MN}$$

The fundamental equation then reads

$$1.116 \frac{dN_s}{d\varphi} + 0.116 N_s + 0.590\,27 = 0$$

The solution of the homogeneous equation yields $N_s = C e^{-0.104\varphi}$,
the particular solution is $\qquad\qquad N_s = -5.088$ MN
and the total solution yields $\qquad\qquad N_s = C e^{-0.104\varphi} - 5.088$ (a)
In the initial state for $t = 0$, after subjecting the member to the force N_t, the

reinforcement takes a force, which is determined from the condition $N_t = N_s + N_c$ and $\varepsilon_s = \varepsilon_c$; this yields

$$N_{s0} = -\frac{\dfrac{N_t}{E_c A_c}}{\dfrac{1}{E_s A_s} + \dfrac{1}{E_c A_c}} = \frac{\dfrac{3.8}{0.036 \times 0.48}}{\dfrac{1}{0.21 \times 0.009\,914} + \dfrac{1}{0.036 \times 0.48}} =$$

$$= -0.409 \text{ MN}$$

$$N_{c0} = -3.8 + 0.409 = -3.391 \text{ MN}$$

At the application of the load ($t = 0$), the reinforcement must take the force N_s; hence, for $\varphi = 0$

$$C\,e^{-0.104\varphi} - 5.088 = -0.409$$

and then

$$C = 4.679 \text{ MN}$$

As $t \to \infty$, the reinforcement takes the force (with reference to Eq. (a))

$$N_s = 4.679^{-0.104 \times 1.98} - 5.088 = -1.279\,8 \text{ MN}$$

Approximate analysis

Equation (3.56a) is arranged into the form (with $N_0 = N_{c0}$)

$$\bar{A}_1''' N_s - \mu n \left(1 + \frac{\varphi}{2}\right) N_t - \mu n \frac{\varphi}{2} N_t(t) + E_s A_s \varepsilon_s = 0$$

$$\bar{A}_1''' = 1 + 0.02 \times 5.8 \left(1 + \frac{1.98}{2}\right) = 1.230\,8$$

$$N_s = \left[-0.02 \times 5.8 \times 3.8 \left(1 + \frac{1.98}{2}\right) - 0.02 \times 5.8 \times 3.391\,\frac{1.98}{2} - \right.$$

$$\left. - 0.21 \times 10^6 \times 0.009\,914 \times 14.22 \times 10^{-5}\right] : 1.230\,8 = -1.269\,6 \text{ MN}$$

Prestressed cross-sections

The losses of prestress caused by creep and shrinkage of concrete can be analysed in accordance with the principles introduced in the preceding Sections. The procedure for the deduction of the equations needed for the analysis will be demonstrated in a cross-section, where the prestressing reinforcement is placed at both the top and bottom surfaces of the member (Fig. 3.7).

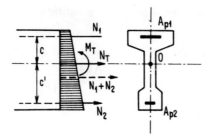

Fig. 3.7 Prestressed cross-section

Calculation by the rate-of-creep theory

Designating the losses of prestress due to creep as ΔN_1 and ΔN_2, for the forces acting in the prestressed reinforcement at an instant t, we may write

$$N_1(t) = N_{01} - \Delta N_1(t); \quad N_2(t) = N_{02} - \Delta N_2(t)$$

where N_{01} and N_{02} are the prestressing forces at the instant when prestressing has been applied to the member (the instantaneous changes of prestressing are already included in the forces N_{01} and N_{02}).

The force by which the prestressed elements act on concrete, is

$$N_p(t) = N_1(t) + N_2(t) = N_c(t)$$

and its point of action is situated at a distance c from the centroid of the concrete cross-section. In addition, if the concrete is subjected to a force N_t and to a bending moment M_t, and if the concrete is subjected to shrinkage, the following strain conditions can be written:

at the axis of the top reinforcement

$$\mu_1 n \left[\frac{d(\Delta N_1 + \Delta N_2)}{d\varphi} - (N_{01} - \Delta N_1) - (N_{02} - \Delta N_2) \right] -$$

$$- \mu_1 n \frac{c}{i^2} \left[\frac{d(\Delta N_1 c - \Delta N_2 c')}{d\varphi} - (N_{01} - \Delta N_1) c + (N_{02} - \Delta N_2) c' \right] -$$

$$- E_s A_{p1} \frac{\varepsilon_{s\infty}}{\varphi_\infty} + \mu_1 n N_t - \mu_1 n \frac{c}{i^2} M_t = - \frac{d \Delta N_1}{d\varphi} \tag{3.57a}$$

at the axis of the bottom reinforcement

$$\mu_1 n \left[\frac{d(\Delta N_1 + \Delta N_2)}{d\varphi} - (N_{01} - \Delta N_1) - (N_{02} - \Delta N_2) \right] -$$

$$- \mu_2 n \frac{c'}{i^2} \left[\frac{d(\Delta N_1 c - \Delta N_2 c')}{d\varphi} - (N_{01} - \Delta N_1) c + (N_{02} - \Delta N_2) c' \right] -$$

$$- E_s A_{p2} \frac{\varepsilon_{s\infty}}{\varphi_\infty} + \mu_2 n N_t + \mu_2 n M_t \frac{c'}{i^2} = - \frac{d \Delta N_2}{d\varphi} \tag{3.57b}$$

Simultaneously, these conditions represent the relationship for the determination of the unknown losses of prestress ΔN_1 and ΔN_2, and they may be modified thus:

$$B_1 \frac{d\Delta N_1}{d\varphi} + B_2 \frac{dN_2}{d\varphi} + B_3 \Delta N_1 + B_2 \Delta N_2 - B_3 N_{01} - B_2 N_{02} +$$

$$+ \mu_1 n \left(N_t - \frac{c}{i^2} M_t \right) - E_s A_{p1} \frac{\varepsilon_{s\infty}}{\varphi_\infty} = 0 \qquad (3.58a)$$

$$B_1' \frac{d\Delta N_2}{d\varphi} + B_2' \frac{d\Delta N_1}{d\varphi} + B_3' \Delta N_2 + B_2' \Delta N_1 - B_3' N_{02} - B_2' N_{01} +$$

$$+ \mu_2 n \left(N_t + \frac{c}{i^2} M_t \right) - E_s A_{p2} \frac{\varepsilon_{s\infty}}{\varphi_\infty} = 0 \qquad (3.58b)$$

$$B_1 = 1 + \mu_1 n \left(1 + \frac{c^2}{i^2} \right); \quad B_2 = \mu_1 n \left(1 - \frac{cc'}{i^2} \right); \quad B_3 = \mu_1 n \left(1 + \frac{c^2}{i^2} \right)$$

$$\mu_1 = \frac{A_{p1}}{A_c}; \quad \mu_2 = \frac{A_{p2}}{A_c}; \quad n = \frac{E_p}{E_c}$$

where c and c' are the distances of the axes of the prestressing reinforcement from the centroid of the concrete section; i is the radius of inertia of the concrete section. The coefficients B_1' to B_3' are found by substituting c for c' and μ_1 for μ_2.

Modification of the equations using the mean stress value

Using the relationship

$$\varepsilon(t) = \frac{\sigma(t)}{E} + \frac{\sigma_0 + \sigma(t)}{E} \frac{\varphi(t)}{2}$$

and applying the designations of the quantities from the previous examples, the strain-compatibility conditions will have the form:
at the axis of the top reinforcement:

$$\left\{ \mu_1 n [-(N_{01} - \Delta N_1) - (N_{02} - \Delta N_2)] - \right.$$

$$- \mu_1 n \frac{c}{i^2} [(N_{01} - \Delta N_1) c - (N_{02} - \Delta N_2) c'] + \mu_1 n N_t -$$

$$\left. - \mu_1 n \frac{c}{i^2} M_t \right\} \left(1 + \frac{\varphi}{2} \right) - E_s A_{p1} \varepsilon_s + \mu_1 n \left[-N_{01} - N_{02} - \right.$$

$$\left. - \frac{c}{i^2} (N_{01} c - N_{02} c') + \mu_1 n N_t - \mu_1 n \frac{c}{i^2} M_t \right] \frac{\varphi}{2} = -\Delta N_1 \qquad (3.59a)$$

at the axis of the bottom reinforcement:

161

$$\left\{\mu_2 n\left[-(N_{01} - \Delta N_1) - (N_{02} - \Delta N_2)\right] + \right.$$

$$+ \mu_2 n \frac{c}{i^2}\left[(N_{01} - \Delta N_1)c - (N_{02} - \Delta N_2)c'\right] + \mu_2 n N_t +$$

$$+ \mu_2 n \frac{c'}{i^2} M_t\bigg\}\left(1 + \frac{\varphi}{2}\right) - E_s A_{p2}\varepsilon_s + \mu_2 n\left[-N_{01} - N_{02} + \right.$$

$$+ \frac{c'}{i^2}(N_{01}c - N_{02}c') + \mu_2 n N_t + \mu_2 n \frac{c'}{i^2} M_t\bigg]\frac{\varphi}{2} = -\Delta N_2 \qquad (3.59b)$$

where ε_s designates shrinkage in the considered time interval.

For the analysis, the conditions may be arranged into more convenient form:

$$(B_1 \Delta N_1 + B_2 \Delta N_2)\left(1 + \frac{\varphi}{2}\right) -$$

$$- \left[B_3 N_{01} + B_2 N_{02} + \mu_1 n\left(N_t - \frac{c}{i^2} M_t\right)\right]\left(1 + \frac{\varphi}{2}\right) -$$

$$- \left[B_3 N_{01} + B_2 N_{02} - \mu_1 n\left(N_t - \frac{c}{i^2} M_t\right)\right]\frac{\varphi}{2} - E_s A_{p1}\varepsilon_s = 0 \qquad (3.60a)$$

$$(B_1' \Delta N_2 + B_2' \Delta N_1)\left(1 + \frac{\varphi}{2}\right) -$$

$$- \left[B_3' N_{02} + B_2' N_{01} - \mu_2 n\left(N_t + \frac{c'}{i^2} M_t\right)\right]\left(1 + \frac{\varphi}{2}\right) -$$

$$- \left[B_3' N_{02} + B_2' N_{01} - \mu_2 n\left(N_t + \frac{c'}{i^2} M_t\right)\right]\frac{\varphi}{2} - E_s A_{p2}\varepsilon_s = 0 \qquad (3.60b)$$

The constants B_1 to B_3 and B_1' to B_3' are given in the preceding Section.

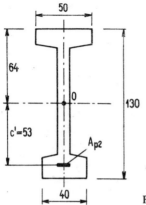

Fig. 3.8 Cross-section of the analysed prestressed member

Example

The cross-section portrayed in Fig. 3.8 is prestressed by an initial force $N_{02} = 2.1$ MN and subjected to a bending moment $M_t = 1.3$ MN m; the section is prestressed and subjected to the load simultaneously 28 days after loading; the loss of prestressing force for $t \to \infty$ is to be determined.

$$A_c = 0.286 \ m^2; \quad i^2 = 0.223 \ m^2; \quad \mu_2 = 0.007 \ 34; \quad n = 5.28;$$

$$A_{p2} = 0.002 \ 1 \ m^2; \quad E_s = 0.21 \times 10^6 \ MPa; \quad \varepsilon_{s\infty} = 28 \times 10^{-5};$$

$$\varphi_\infty = 3.9; \quad c' = 0.53 \ m;$$

φ for the interval $\langle 28 \ \text{days}, \infty \rangle = 1.98$;

ε_s (for the same interval) $= 14.22 \times 10^{-5}$;

$$B_1' = 1 + 0.007 \ 34 \times 5.28 \left(1 + \frac{0.53^2}{0.223}\right) = 1.087 \ 9$$

$$B_3' = 0.007 \ 34 \times 5.28 \left(1 + \frac{0.53^2}{0.223}\right) = 0.087 \ 9$$

More accurate analysis

Substituting into Eqs. (3.58b) with $\Delta N_1 = N_{01} = N_t = 0$ we can write

$$1.087 \ 9 \frac{d \Delta N_2}{d\varphi} + 0.087 \ 9 \Delta N_2 - 0.087 \ 9 \times 2.1 +$$

$$+ 0.007 \ 34 \times 5.28 \frac{0.53}{0.223} 1.3 - 0.21 \times 10^6 \times 0.002 \ 1 \frac{28.0}{3.9} 10^{-5} = 0$$

and hence

$$1.087 \ 9 \frac{d \Delta N_2}{d\varphi} + 0.087 \ 9 \Delta N_2 - 0.096 \ 510 = 0$$

The solution yields

$$\Delta N_2 = C \, e^{-0.0808\varphi} + 1.097 \ 95$$

and the constant C is determined from the condition that at $\varphi = 0$, $\Delta N_2 = 0$; then $C = -1.097 \ 95$ and $\Delta N_2 = 1.097 \ 95(1 - e^{-0.0808\varphi})$. The final loss of prestress is

$$\Delta N_2 = 1.097 \ 95(1 - e^{-0.0808 \times 1.98}) = 0.162 \ 3 \ MN$$

Approximate analysis

Equation (3.60b), after substitution, is

$$1.087 \ 9 \left(1 + \frac{1.98}{2}\right) \Delta N_2 - 0.087 \ 9 \left(1 + \frac{1.98}{2}\right) 2.1 + 0.007 \ 34 \times 5.28$$

$$\frac{0.53}{0.223} 1.3 \left(1 + \frac{1.98}{2}\right) -$$

$$- \left(0.087\,9 \times 2.1 - 0.007\,34 \times 5.28 \frac{0.53}{0.223} 1.3\right)$$

$$\frac{1.98}{2} - 0.21 \times 10^6 \times 0.002\,1 \times 14.22 \times 10^{-5} = 0$$

From this

$$\Delta N_2 = 0.118\,2 \text{ MN}$$

The difference between both these solutions is large; however, if the stresses in the outermost fibres of the concrete cross-section are calculated from both these analyses, the difference between the "accurate" and the approximate solution is much less:

The accurate solution yields the stress in the top fibres

$$\sigma_u = 9.51 \text{ MPa},$$

in the bottom fibres

$$\sigma_b = -3.95 \text{ MPa}.$$

The approximate solution yields

$$\sigma_u = 9.43 \text{ MPa},$$

$$\sigma_b = -4.34 \text{ MPa}.$$

4

Analysis of structures
with fully acting cross-sections

Creep and shrinkage of concrete may exert a very significant effect on the behaviour of structures. The most common manifestation of creep, which is respected in the design of concrete structures, is the growth of deformations. Some European bridges of large spans, made of prestressed concrete and constructed by segmental cantilevering with hinges at mid-span, may serve as an example. The deflections of these bridges have been known to be so large as to affect their serviceability, [4.1].

In concrete columns, compressed walls and thin shells, creep significantly augments not only the deflections, but also the internal forces in the structure which, in these cases, depend on the change of shape of the structure. This may lead to a long-term instability of the structure (the magnitude of the deflections does not stabilize, and the deflections continue to grow in time); the admissible load in such instances may be several times less than the load at short-term loading.

Long continuous bridges (especially of prestressed concrete) are nowadays constructed by employing many advanced methods, such as, for example, cantilevering, slip formwork, segmental construction slip-out, erection by longitudinal elements, etc. Various operations used in construction, such as addition of segments, prestressing, changing of support boundary conditions, application or removal of construction loads and prescribed displacements, are the causes of unavoidable changes of stress. It is characteristic for all these methods that the structure changes the system of its static behaviour during the construction process (usually several times), that the stress changes gradually (including the prestressing), and that usually the structure becomes non-homogeneous owing to the different ages of concrete or due to the ambient conditions. Variations of temperature should also be included among these factors. For these reasons, creep of concrete and/or shrinkage become significant, and their correct analysis is a condition of a safe and economical design.

The analysis of the effect of creep on the behaviour of concrete structures involves some inaccuracies with regard to reality. They may be divided into two groups: the first group results from an insufficient knowledge of the laws governing

165

the physical substance of creep and of their mathematical expression. The restriction of these errors is a problem for materials research. The other group of inaccuracies is the consequence of simplifications which have to be introduced into the calculations, if they are to be carried out with the available mathematical apparatus.

Most of the techniques applied to these analyses (within the range of stresses appearing under current service conditions of a structure) have the character of a compromise between the endeavour to achieve an exact formulation, together with an admissible simplification of the resulting formulae, and the possibility of mathematical operations performed with accessible computing facilities. Lately, the reliability of the results has improved — thanks to the computers, and the recent rheological knowledge can be considered in the analyses.

In the course of time, particularly within the last two decades, many different methods of analysis of the effects of creep and shrinkage on concrete structures have been developed. These methods may be divided, like in other areas of the theory of structures, into numerical and analytical techniques.

The numerical methods either solve the governing equations by approximate methods, or, more often in recent years, they start at the very beginning from the assumption of an appropriate space and time-discretization of the continuum and from its time-dependent behaviour. At the same time, there appear current trends to employ computers: to obtain numerical results (which correspond to the character of computer) by means of programs based either on the repetition of analysis or on successive approximations, or on a step-by-step analysis in relatively short (time and space) intervals on the basis of comparatively simple and verified algorithms. For the absolute majority of cases, the routine solution of a set of linear algebraic equations represents the principal part of the analysis.

This is quite different from the analytical methods of analysis; here, the formulation of the problem usually leads to systems of governing differential equations, where a general solution is to be found in the first instance and the desired solution of the problem is established by the application of the respective boundary and initial conditions. The advantage of this procedure, above all, is in the general character of the results, in their analytical expression which, in some cases, can be formulated. This allows correct opinions to be formed about the character and course of the investigated phenomenon. The disadvantage is the unavoidable repetition of the entire procedure, when parameters are changed, and the difficult algorithmization of the analysis. A closed form solution is obtainable only exceptionally, and hence, frequently, in the last resort, the analytically set-up differential equations have to be solved by some suitable numerical method.

The choice of the method is influenced not only by the character of the problem (importance and complexity of a structure, stage of solution, preliminary or final design), but also by the availability of suitable computer facilities.

The most commonly applied theory for concrete structures is the linear aging

166

viscoelastic theory for which the superposition theorem applies. The only numerical complication is the problem of storing the past stress history. Expansion of the creep function in a finite Dirichlet series (see Sect. 2.1.3), however, simplifies the storage problem and renders possible the numerical solution of complex concrete structures. Thus, the difficulty of computing sums involving the complete stress history at each time interval is avoided [4.3].

The problems related to creep of concrete are important in line structures as well as in structures behaving spatially. The classification of a structure in one of the two groups depends on its final state, but also on the sequence and character of the individual construction stages. For example, a straight bridge of symmetrical double-cell box-section may be represented by a continuous beam for the analysis of the dead load and prestressing effects if it is constructed in its full width; however, this bridge had to be analysed as a three-dimensional system if it were to be constructed gradually along its longitudinal halves (see e.g. Fig. 4.12). More complex systems, such as shells, are three-dimensional by their character and they have to be considered as such in the analysis.

4.1 Analysis of structural systems consisting of bar members

The present discussion deals with structures of fully acting cross-sections composed of bar members. The structure may be non-homogeneous, i.e., its individual components may be made of concretes of various ages, or of different material altogether, e.g. steel. In particular, this group includes frame structures with all their variants (statically determinate and indeterminate beams, plane frames, arches and space frames); a factor common to all these structures, however, is that they satisfy the assumption of long members in relation to the dimensions of their cross-sections.

In the cross-section of a structural member, the following internal forces act:

$$\{S\}^T = N, M_y, M_z, M_t, Q_z, Q_y \tag{4.1}$$

where N is the normal force;

    M_y, M_z are the bending moments;

    M_t is the torsional moment;

    Q_z, Q_y are the shear forces in the local coordinate system of the member, whose x-axis corresponds to the tangent to the median line s of the member (Fig. 4.1).

Besides the internal forces, described by Eq. (4.1), other quantities have to be considered in a thin-walled member: a bi-moment, and, in the case of a deformable cross-section, also distortional forces deforming the cross-section, characterized

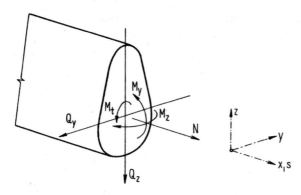

Fig. 4.1 Section through a member element showing the internal forces at the cross-section; the co-ordinate system used

by a transverse bi-moment. This, however, would mean only an extension of the discussion, which concerns the members of solid section, by further terms. The principle would remain the same.

4.1.1 Analysis based on the rate-of-creep theory

The rate-of-creep theory is unavoidably deficient in one respect, namely that it is unable to regard the loading history. It completely neglects the delayed elastic deformation, which may lead to rather large deviations from the actual behaviour of the structure, especially when it is non-homogeneous, or when an older concrete is loaded.

The structural analysis based on this theory, however, allows the results to be expressed analytically and to obtain a clear idea about the development of the investigated phenomena. Hence, it is pertinent to introduce this method in the summary of usable procedures for calculations. Moreover, on the basis of the results obtained according to the rate-of-creep theory and using the subsequently discussed simple method of an effective time of action of redundants (Sect. 4.1.3), a general form of creep function can be considered and, in this way, reasonably accurate and practical results can be obtained.

In the case of a homogeneous cross-section (i.e., neglecting, for example, the effect of the reinforcement), the unrestrained shrinkage of concrete in a time interval $\langle \tau, t \rangle$ merely induces longitudinal strains of the members, viz.

$$\varepsilon_s(t) = \frac{\varepsilon_{s\infty}}{\varphi_\infty} \left[\varphi(t) - \varphi(\tau) \right] \tag{4.2}$$

where $\varepsilon_{s\infty}$ is the final value of shrinkage (compare with Eq. (2.28)).

Defining the vector $\{\omega\}$ as

$$\{\omega\}^T = -\frac{\varepsilon_{s\infty}}{\varphi_\infty}, 0, 0, 0, 0, 0 \tag{4.3}$$

168

the following applies

$$\{\varepsilon_s(t)\} = \{\omega\}[\varphi(t) - \varphi(\tau)] \tag{4.4}$$

If the internal forces $\{S\}$ are variable in time, the total strains (elastic and irreversible due to creep and shrinkage, and strains induced by temperature) are given by the relation

$$\{\varepsilon(t)\} = [H]\left(\{S(t)\} + \int_{t_0}^{t} \{S(\tau)\}\frac{d\varphi(\tau)}{d\tau}\,d\tau\right) + \{\varepsilon_s(t)\} + \{\varepsilon_T(t)\} \tag{4.5}$$

where $\{\varepsilon_T(t)\}$ are the strains induced by temperature, which are generally variable in time (also see (2.123)), and

$$[H] = \begin{bmatrix} \dfrac{1}{EA}, & 0, & 0, & 0, & 0, & 0 \\[2mm] 0, & \dfrac{1}{EI_y}, & 0, & 0, & 0, & 0 \\[2mm] 0, & 0, & \dfrac{1}{EI_z}, & 0, & 0, & 0 \\[2mm] 0, & 0, & 0, & \dfrac{1}{GI_t}, & 0, & 0 \\[2mm] 0, & 0, & 0, & 0, & \dfrac{\gamma_z}{GA}, & 0 \\[2mm] 0, & 0, & 0, & 0, & 0, & \dfrac{\gamma_y}{GA} \end{bmatrix}$$

is the matrix of cross-section flexibility constants (A is the cross-sectional area, I_y and I_z the second moments of area of the cross-section, GI_t the torsional stiffness, and γ_z, γ_y are the coefficients expressing influence of the distribution of shear stresses on deformation).

For the time-dependent variation of strain we obtain

$$\frac{d}{dt}\{\varepsilon(t)\} = [H]\left(\frac{d}{dt}\{S(t)\} + \{S(t)\}\frac{d\varphi(t)}{dt}\right) +$$

$$+ \{\omega\}\frac{d\varphi(t)}{dt} + \frac{d}{dt}\{\varepsilon_T(t)\} \tag{4.6}$$

Applying the principle of virtual works, we can write for the time-dependent variation of displacement $\left(\dfrac{d\delta_i(t)}{dt}\right)$

$$\frac{d\delta_i(t)}{dt} = \int_{\Omega}\{\bar{S}_i\}^T\frac{d}{dt}\{\varepsilon(t)\}\,ds =$$

$$= \int_\Omega \{\bar{S}_i\}^T \left[[H] \left(\frac{d}{dt} \{S(t)\} + \{S(t)\} \frac{\partial \varphi(s, t)}{\partial t} \right) + \right.$$

$$\left. + \{\omega\} \frac{\partial \varphi(s, t)}{\partial t} + \frac{\partial}{\partial t} \{\varepsilon_T(s, t)\} \right] ds \tag{4.7}$$

where i designates the point, the direction and the sense of the desired displacement at this point, $\{\bar{S}_i\}$ are the internal forces induced by the external virtual quantity $\bar{X}_i = 1$ acting in the direction i, Ω designates integration over the entire structure, s is the coordinate measured along the median line of the structural members (Fig. 4.1). The function $\varphi(s, t)$ also depends on the coordinate s in the case of non-homogeneous structure.

In the analysis of statically indeterminate frame structures with consideration of the effect of creep and shrinkage by means of the force method, the procedure can be similar to the case of elastic structures.

Let us consider a k-fold statically indeterminate structure which is transformed into a primary statically determinate structure by releasing k restraints.[*)] Redundants $X(t)$, variable in time, act at the points of the released restraints. The cross-sections of the members of the primary structure loaded by external forces are stressed by the internal forces $\{\mathfrak{S}\}$. Under the action of $\bar{X}_i = 1$ or $X_j = 1$, internal forces $\{\bar{S}_i\}$ or $\{S_j\}$ develop in the primary system. The stress of the actual statically indeterminate structure is characterized by the internal forces $\{S\}$ for which

$$\{S\} = \{\mathfrak{S}\} + \sum_{j=1}^{k} \{S_j\} X_j \tag{4.8}$$

The compatibility of the structure at the points of released restraints must be maintained throughout the period of action of the statically indeterminate structure. This means that the resulting time-variation of displacement on the primary system at the points of released restraints will be zero (or its time-distribution will be prescribed by a function describing the development of an external intervention in the structure, for example, a change of position of the supports).

Consequently, the following equation must hold for the time-dependent variation of displacement in the sense of the i-th redundant:

$$\frac{d\delta_i(t)}{dt} = \int_\Omega \{\bar{S}_i\}^T \left[[H] \left[\frac{\partial}{\partial t} \left(\{\mathfrak{S}\} + \sum_{j=1}^{k} \{S_j\} X_j \right) + \right. \right.$$

$$\left. + \left(\{\mathfrak{S}\} + \sum_{j=1}^{k} \{S_j\} X_j \right) \frac{\partial \varphi(s, t)}{\partial t} \right] + \{\omega\} \frac{\partial \varphi(s, t)}{\partial t} +$$

$$\left. + \frac{\partial}{\partial t} \{\varepsilon_T(s, t)\} \right] ds = \frac{d\delta_{pi}(t)}{dt} \tag{4.9}$$

This equation has been obtained by substituting Eq. (4.8) into Eq. (4.7), where

---

[*)] A primary statically indeterminate system may be chosen as well, if this is advantageous.

$\delta_{pi}(t)$ is the time-dependent distribution of the possible external intervention at the point of the i-th restraint. If equations of the type (4.9) are written for all the k displacements at the points of released restraints on the primary system, a set of differential equations is obtained of the form

$$[\delta_{ij}] \left\{ \frac{dX_j(t)}{dt} \right\} + [_\varphi\delta_{ij}] \{X_j(t)\} + \{P_i(t)\} = 0 \qquad (4.10)$$

where the elements of the flexibility matrix are

$$\delta_{ij} = \int_\Omega \{\bar{S}_i\}^T [H] \{S_j\} \, ds \qquad (4.11)$$

$$_\varphi\delta_{ij} = \int_\Omega \{\bar{S}_i\}^T [H] \{S_j\} \frac{\partial \varphi(s, t)}{\partial t} \, ds \qquad (4.12)$$

The vectors $\{X_j(t)\}$ and $\left\{ \frac{dX_j(t)}{dt} \right\}$ are the redundant vectors and their derivatives, respectively. The vector $\{P_i(t)\}$ describes the effects of the external loading of the structure, the shrinkage and the external interventions; it follows that

$$P_i(t) = \int_\Omega \{\bar{S}_i\}^T \left[[H] \left\{ \frac{\partial}{\partial t} \{\mathfrak{S}(s, t)\} + \{\mathfrak{S}(s, t)\} \frac{\partial \varphi(s, t)}{\partial t} \right\} + \right.$$

$$\left. + \{\omega\} \frac{\partial \varphi(s, t)}{\partial t} + \frac{\partial}{\partial t} \{\varepsilon_T(s, t)\} \right] ds - \frac{d\delta_{pi}(t)}{dt} \qquad (4.13)$$

These general relationships are much simplified, if the analysed structure is not a three-dimensional framework and if the effects of some components of the internal forces can be neglected. Thus, for example, for a plane frame, if only the effects of the bending moments are considered, the elements of the flexibility matrices, expressed in a general form by Eq. (4.11) and (4.12), are simplified to the familiar form

$$\delta_{ij} = \int_\Omega \frac{\bar{M}_i(s) \, M_j(s)}{E I(s)} \, ds \qquad (4.11)^*$$

$$_\varphi\delta_{ij} = \int_\Omega \frac{\bar{M}_i(s) \, M_j(s)}{E I(s)} \frac{\partial \varphi(s, t)}{\partial t} \, ds \qquad (4.12)^*$$

where $\bar{M}_i$ and M_j are the bending moments in the primary system induced by the action of $\bar{X}_i = 1$ and $X_j = 1$, respectively. If the structure is arranged so that shrinkage does not influence the flexural stresses, and if the deformation of the structure is not influenced by any interventions, then Eq. (4.13) acquires the simple form

$$P_i(t) = \int_\Omega \frac{\bar{M}_i(s)}{E I(s)} \left[\frac{\partial \mathfrak{M}(s, t)}{\partial t} + \mathfrak{M}(s, t) \frac{\partial \varphi(s, t)}{\partial t} \right] ds \qquad (4.13a)$$

where $\mathfrak{M}$ is the bending moment on the primary system.

171

The long-term external loading is usually constant with time. In addition, this assumption is currently accepted for structures made of prestressed concrete, although their long-term loading also depends on the magnitude of prestressing, which, however, diminishes due to the losses induced by creep and shrinkage.

In a structure subjected to a loading which is constant with time, the internal forces $\{\mathfrak{S}\}$ in the primary system are also independent of time; hence, in this case, Eqs. (4.13) and (4.13a) are simplified to the form

$$P_i(t) = \int_{\Omega} \{\bar{S}_i\} \left[([H]\{\mathfrak{S}(s)\} + \{\omega\}) \frac{\partial \varphi(s,t)}{\partial t} + \right.$$

$$\left. + \frac{\partial}{\partial t} \{\varepsilon_T(s,t)\} \right] ds - \frac{d\delta_{pi}(t)}{dt} \tag{4.$\overline{13}$}$$

and

$$P_i(t) = \int_{\Omega} \frac{\bar{M}_i(s)\,\mathfrak{M}(s)}{EI(s)} \frac{\partial \varphi(s,t)}{\partial t} ds \tag{4.13a)*}$$

4.1.1.1 *Homogeneous structures*

Equation (4.10) represents a system of k non-homogeneous differential equations of the first order for redundants $X(t)$; the coefficients of the equations are given by the flexibility matrices (Eqs. 4.11) and (4.12)) and the right-hand side is expressed by Eq. (4.13). The analytical solution of this system, i.e. the general integral, is simple, if the structure is homogeneous (of the same modulus of elasticity and the same age) and subjected in time to a constant loading.

In this case of a homogeneous structure, a single creep function holds good for the entire structure, and Eq. (4.12) can be written

$$[_\varphi \delta_{ij}] = [\delta_{ij}] \frac{d\varphi(t)}{dt} \tag{4.14}$$

and the above system of differential equations may be modified to the form

$$[\delta_{ij}] \left(\left\{ \frac{dX_j(t)}{dt} \right\} + \frac{d\varphi(t)}{dt} \{X_j(t)\} \right) + \frac{d\varphi(t)}{dt} \{_0 P_i\} = 0 \tag{4.15}$$

If the structure is not affected by external interventions and if the effects of shrinkage and temperature are not considered, then $\{_0 P_i\}$ is

$$_0 P_i = \int_{\Omega} \{\bar{S}_i\}^T [H] \{\mathfrak{S}\} ds \tag{4.16}$$

Equation (4.15) is satisfied by the solution

$$X_j(t) = A_j(1 - e^{-\varphi} + C_j) \tag{4.17}$$

Substituting Eq. (4.17) into Eq. (4.15), we obtain

$$[\delta_{ij}]\{A_j(1 + C_j)\} + \{_0P_i\} = 0 \tag{4.18}$$

The system of governing linear algebraic equations of the force method for an *elastic* structure at the same loading (with redundants $_eX_j$) would be

$$[\delta_{ij}]\{_eX_j\} + \{_0P_i\} = 0 \tag{4.19}$$

For the sake of clarity and for a comparison with the results of the analysis of an elastic structure, one may begin from the analogy of the systems given by Eqs. (4.18) and (4.19). We obtain

$$A_j(1 + C_j) = {_eX_j} \tag{4.20}$$

hence

$$A_j = \frac{_eX_j}{1 + C_j} \tag{4.21}$$

Substituting Eq. (4.21) into Eq. (4.17) we obtain

$$X_j(t) = \frac{_eX_j}{1 + C_j}(1 - e^{-\varphi} + C_j) \tag{4.22}$$

where C_j is a constant which is dependent on the stress state of the structure at the commencement of the investigated process.

The initial condition for the distribution of the time-variable values of the redundants $X_j(t)$ is formed by their values $(_0X_j)$ at the commencement of the action of the structure in the given static system (i.e. at time $t = t_0$). Hence, it follows

$$X_j(t_0) = {_0X_j} \tag{4.23}$$

Substituting Eq. (4.23) into Eq. (4.22) for time t_0, we obtain the value of the constant

$$C_j = \frac{_eX_j[1 - e^{-\varphi(t_0)}] - {_0X_j}}{_0X_j - {_eX_j}} \tag{4.24}$$

Substitution of Eq. (4.24) into Eq. (4.22) yields the resulting relationship of the time-dependent distribution of the redundants

$$X_j(t) = (_0X_j - {_eX_j})e^{-[\varphi(t) - \varphi(t_0)]} + {_eX_j} \tag{4.25}$$

Several possibilities may occur under actual conditions:

(a) The structure is cast on a centering, which is removed at time t_0, and the structure starts to act immediately in the final static system. At this instant t_0, the redundants equal the values of the redundants appertaining to an elastic structure, i.e. the condition of Eq. (4.23) is

$$X_j(t_0) = {_0X_j} = {_eX_j} \tag{4.26}$$

173

For the time-dependent distribution of the redundant quantities $X_j(t)$, Eq. (4.25) becomes

$$X_j(t) = {}_eX_j \tag{4.27}$$

which means that the redundants X_j are constant with time, and are equal to these quantities on an elastic structure throughout the time-period; consequently, the stress state of the structure does not change at all due to the effect of creep. However, the deformations of the structure increase in time.

(b) Parts of the structure are connected immediately after casting and de-centering so that they act in a more complex static system. If a part of the structure, prior to its incorporation into the whole, is considered as the primary system, in which the forces developing at the joints are considered as the redundants, and if it is theoretically assumed that the connection of the parts into a whole occurs at time $t_0 = 0$, then the following initial condition applies

$$X_j(t_0) = X_j(0) = {}_0X_j = 0 \tag{4.28}$$

Since, simultaneously, $\varphi(0) = 0$, we obtain, in agreement with Eq. (4.24)

$$C_j = 0 \tag{4.29}$$

Hence, the time-distribution of the redundants, with consideration of Eq. (4.22), or, directly, according to Eq. (4.25), is given by the equation

$$X_j(t) = {}_eX_j(1 - e^{-\varphi(t)}) \tag{4.30}$$

and the limiting value for $t \to \infty$ is

$$X_j(\infty) = {}_eX_j(1 - e^{-\varphi(\infty)}) \tag{4.31}$$

(c) Parts of the structure are not connected into the final static system immediately, as in the preceding case, but only after a certain time t_0. Then, the following initial condition applies

$$X_j(t_0) = {}_0X_j = 0 \tag{4.32}$$

We can write, in accordance with Eq. (4.24),

$$C_j = e^{-\varphi(t_0)} - 1 \tag{4.33}$$

After substituting Eq. (4.33) into Eq. (4.22), or, in accordance with Eq. (4.25), the time-distribution of the redundants is

$$X_j(t) = {}_eX_j(1 - e^{-[\varphi(t) - \varphi(t_0)]}) \tag{4.34}$$

and its limiting value for $t \to \infty$ is

$$X_j(\infty) = {}_eX_j(1 - e^{-[\varphi(\infty) - \varphi(t_0)]}) \tag{4.35}$$

174

The effect of enforced deformations (constructional interventions, differential settlement of supports, deformations imposed by jacks for the purpose of rectifying the state of a structure, etc.) on a homogeneous structure can be investigated in the same manner.

In this case, Eq. (4.13) is simplified to

$$P_i(t) = -\frac{d\delta_{pi}(t)}{dt} \tag{4.36}$$

and the system of differential equations acquires the form

$$[\delta_{ij}] \left\{ \frac{dX_j(t)}{dt} + \frac{d\varphi(t)}{dt} X_j(t) \right\} - \left\{ \frac{d\delta_{pi}(t)}{dt} \right\} = 0 \tag{4.37}$$

This system is satisfied by the solution

$$X_j(t) = K_j e^{-\varphi(t)} + \bar{X}_j(t) \tag{4.38}$$

where $\bar{X}_j(t)$ is the particular solution.

(A) The most common case is a sudden intervention into the geometry of the structure (for example, settlement of the supports), which occurs at time t_0 and whose value does not vary any more. Then

$$\frac{d\delta_{pi}(t)}{dt} = 0 \tag{4.39}$$

and also

$$\bar{X}_j(t) = 0 \tag{4.40}$$

Hence, the time-dependent distribution of the redundants is

$$X_j(t) = K_j e^{-\varphi(t)} \tag{4.41}$$

The constant K is determined from the initial condition, i.e. at instant t_0, the structure deforms elastically and redundants $_eX$ develop in it. Consequently, it follows

$$X_j(t_0) = {}_eX_j \tag{4.42}$$

Substituting into Eq. (4.41), we obtain for the constant

$$K_j = {}_eX_j e^{\varphi(t_0)} \tag{4.43}$$

and, in accordance with Eq. (4.41), the time-distribution of the redundants is

$$X_j(t) = {}_eX_j e^{-[\varphi(t) - \varphi(t_0)]} \tag{4.44}$$

Substituting Eq. (4.44) into Eq. (4.8) and considering that $\{\mathfrak{S}\} = 0$ for this case, we obtain

175

$$\{S\} = \sum_{j=1}^{k} \{S_j\}_e X_j e^{-[\varphi(t) - \varphi(t_0)]} \tag{4.45}$$

i.e. the time-dependent distribution of the stress of the entire structure is the same as that of the redundants.

Substituting Eq. (4.45) into Eq. (4.6), we obtain

$$\frac{d}{dt}\{\varepsilon(t)\} = [H]\left(\frac{d}{dt}\{S(t)\} + \{S(t)\}\frac{d\varphi(t)}{dt}\right) =$$

$$= [H]\sum_{j=1}^{k}\left(-\{S_j\}_e X_j e^{-[\varphi(t) - \varphi(t_0)]}\frac{d\varphi(t)}{dt} + \right. \tag{4.46}$$

$$\left. + \{S_j\}_e X_j e^{-[\varphi(t) - \varphi(t_0)]}\frac{d\varphi(t)}{dt}\right) = 0$$

This means that, when the constant value of forced deformation at the points of action of the redundants in a homogeneous structure is maintained, the variation in time of the deformations of all parts of the structure is zero.

(B) An interesting case is that, when the distribution of the intervention follows that of the creep function of concrete. Then

$$\frac{d\delta_{pi}(t)}{dt} = \frac{\delta_{pi}(\infty)}{\varphi(\infty)}\frac{d\varphi(t)}{dt} \tag{4.47}$$

where $\delta_{pi}(\infty)$ is the final value of the enforced deformation in the sense of the i-th redundant, assuming that the process developed from instant $t = 0$.

By introducing Eq. (4.47) into Eq. (4.37), we obtain a system of differential equations

$$[\delta_{ij}]\left\{\frac{dX_j(t)}{dt} + \frac{d\varphi(t)}{dt}X_j(t)\right\} - \frac{d\varphi(t)}{dt}\left\{\frac{\delta_{pi}(\infty)}{\varphi(\infty)}\right\} = 0 \tag{4.48}$$

whose solution may be conveniently assumed to be of the form (compare with Eqs. (4.15) and (4.18))

$$X_j(t) = \frac{B_j}{\varphi(\infty)}(1 - e^{-\varphi} + R_j) \tag{4.49}$$

Introducing Eq. (4.49) into Eq. (4.48), we obtain the system

$$[\delta_{ij}]\{B_j(1 + R_j)\} = \{\delta_{pi}(\infty)\} \tag{4.50}$$

which is analogous to the system

$$[\delta_{ij}]\{_e X_j\} = \{\delta_{pi}(\infty)\} \tag{4.51}$$

This is the system of governing equations of the force method for the analysis of an elastic structure with redundants $_e X$, the structure being stressed by the final values of prescribed deformations $\delta_{pi}(\infty)$ at the points of action of the redundants (for example, settlements of supports).

Comparing the systems: Eqs. (4.50) and (4.51), we obtain (compare with Eqs. (4.18) and (4.19))

$$B_j = \frac{_eX_j}{1 + R_j} \tag{4.52}$$

and substituting Eq. (4.52) into Eq. (4.49), we obtain the following relationship for the time-distribution of the redundants

$$X_j(t) = \frac{_eX_j}{(1 + R_j)\,\varphi(\infty)}\,(1 - e^{-\varphi} + R_j) \qquad \cdot \tag{4.53}$$

The initial condition is represented by the values of the redundants $_0X_j$ at the commencement of the investigated interval, i.e., at the instant t_0 when intervention occurs. Consequently,

$$X_j(t_0) = {_0X_j} \tag{4.54}$$

Introducing Eq. (4.54) into Eq. (4.53), the constant R_j can be expressed in relation to the values of the redundants $_eX_j$ of the elastic structure

$$R_j = \frac{_eX_j[1 - e^{-\varphi(t_0)}] - {_0X_j}\varphi(\infty)}{_0X_j\varphi(\infty) - {_eX_j}} \tag{4.55}$$

and Eq. (4.53) acquires the form

$$X_j(t) = \left({_0X_j} - \frac{_eX_j}{\varphi(\infty)} \right) e^{-[\varphi(t) - \varphi(t_0)]} + \frac{_eX_j}{\varphi(\infty)} \tag{4.56}$$

The process begins at instant t_0, when known deformations δ_{pi0} are introduced into the structure at the points of action of the redundants; this induces redundants $_0X_j$ in the structure which represent the initial condition for the distribution of the redundants $X_j(t)$ variable in time; the distribution is generally described by Eq. (4.56). Usually, several possibilities may occur:

(a) at time t_0, the following deformations are introduced into the structure:

$$\delta_{pi0} = \frac{\delta_{pi}(\infty)}{\varphi(\infty)} \tag{4.57}$$

In this particular case, the redundants $_0X_j$ develop in the structure at instant t_0 (compare with Eq. (4.51)), so that

$$_0X_j = \frac{_eX_j}{\varphi(\infty)} \tag{4.58}$$

177

which expresses the initial condition for Eq. (4.56). Introducing Eq. (4.58) into Eq. (4.56), we obtain

$$X_j = \frac{{}_e X_j}{\varphi(\infty)} \tag{4.59}$$

In this case, the above equation means that the redundants are constant in time, and, moreover, independent of the instant t_0 at which the intervention into the structure was carried out.

In consequence, the stress state of the structure does not vary, and the deformations at the points of action of the redundants increase from the value δ_{pio} in agreement with Eq. (4.47), i.e.

$$\delta_{pi}(t) = \delta_{pio} + \frac{\delta_{pi}(\infty)}{\varphi(\infty)} [\varphi(t) - \varphi(t_0)] \tag{4.60}$$

for $t \geq t_0$.

This relation may be written, with reference to Eq. (4.57), in the form

$$\delta_{pi}(t) = \frac{\delta_{pi}(\infty)}{\varphi(\infty)} [1 + \varphi(t) - \varphi(t_0)] \tag{4.61}$$

The time-distribution of these deformations, with varying t_0, is represented in Fig. 4.2.

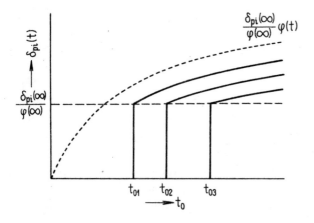

Fig. 4.2 Distribution of deformations in relation to the time of intervention

(b) If deformations of a general magnitude δ_{pio} are introduced into the structure at time t_0, redundants ${}_0 X_j$ develop in the structure at this instant; from Eq. (4.56), we obtain the time-dependent distribution of the quantities redundant in the structure.

(c) If the process of enforcing the deformations at the points of action of the statically indeterminate quantities starts at time t_0 so that the structure is without stress at this instant, i.e. the process starts with zero values of enforced deformations

178

($\delta_{pio} = 0$), then $_0X_j = 0$ and the time-dependent distribution of the statically indeterminate quantities is obtained by Eq. (4.56)

$$X_j(t) = \frac{_eX_j}{\varphi(\infty)}(1 - e^{-[\varphi(t) - \varphi(t_0)]})$$ (4.62)

(d) If the process, as in the preceding case, started with zero values of enforced deformations ($\delta_{pio} = 0$), which then developed (theoretically) from the beginning of the existence of the structure, i.e. from $t_0 = 0$, the time-dependent distribution of the statically indeterminate quantities would result from Eq. (4.62), i.e.

$$X_j(t) = \frac{_eX_j}{\varphi(\infty)}(1 - e^{-\varphi(t)})$$ (4.63)

4.1.1.2 *Non-homogeneous structures*

The analysis of a non-homogeneous structure, possibly subjected also to a load variable in time, can be solved in closed form only for special cases of the creep coefficient, or for special cases of the distribution of loading of the structure in time. This can be accomplished by the solution of a system of differential equations developed from the general procedure which was discussed earlier.

The most common case is a structure whose individual components are cast using concrete of different ages and where loading is constant in time. The system of Eqs. (4.10) may then be solved in closed form, if affinity applies to the creep coefficient $\varphi(s, t)$ of the individual components of the structure, i.e. if these functions are linked by the relation

$$\frac{\partial \varphi_r(s, t)}{\partial t} = \varkappa_r \frac{d\bar{\varphi}(t)}{dt} \quad ^*)$$ (4.64)

(also compare with Eq. (3.13))
where r designates the considered component of the structure;
$\varkappa_r$ is the coefficient of affinity;
$\bar{\varphi}$ is the creep coefficient considered as a reference.

---

*) Equation (4.64) is accurate when the creep coefficient has the form of Eq. (2.8a), i.e.

$$\varphi(t) = \varphi(\infty)(1 - e^{-B_2 t})$$

where B_2 is a constant obtained from tests. However, the assumption of affinity does not apply accurately to the creep coefficient according to Eq. (2.8b)

$$\varphi(t) = \varphi(\infty)\sqrt{1 - e^{-\sqrt{t}}}$$

Equation (4.12) may then be expressed in the form

$$\varphi\delta_{ij} = \frac{d\bar{\varphi}(t)}{dt} \sum_{r=1}^{N} \varkappa_r \int_{\Omega_r} \{\bar{S}_i\}^T [H_r] \{S_j\} \, ds \tag{4.65}$$

where N designates the number of components of the structure of different age; $[H_r]$ is the matrix of the flexibility constants of the cross-section of the r-th part of the structure.

Equation (4.65) yields

$$[_\varphi\delta_{ij}] = \frac{d\bar{\varphi}(t)}{dt} [_x\delta_{ij}] \tag{4.66}$$

where the matrix $[_x\delta_{ij}]$ may also be considered as the flexibility matrix of the elastic structure, whose elastic moduli were reduced by the respective coefficient $\varkappa^*$). If the effects of temperature and shrinkage of concrete are neglected, Eq. (4.10) acquires the form

$$[\delta_{ij}] \left\{ \frac{dX_j(t)}{dt} \right\} + \frac{d\bar{\varphi}(t)}{dt} [_x\delta_{ij}] \{X_j(t)\} + \frac{d\bar{\varphi}(t)}{dt} \{_xP_i\} = 0 \tag{4.67}$$

where again

$$_xP_i = \sum_{r=1}^{N} \varkappa_r \int_{\Omega_r} \{\bar{S}_i\}^T [H] \{\mathfrak{S}\} \, ds \tag{4.68}$$

The solution of the system of non-homogeneous differential equations of the first order (Eq. (4.67)), however, cannot be found as easily as the solution of the system (Eq. (4.15)). The general solution is given by the sum of a general solution of the homogeneous system and of the particular solution corresponding to the right-hand side of the system. The calculation may proceed so that Eq. (4.67) of the first order is transformed by successive arrangements into a single differential equation of the k-th order (k being the degree of static redundancy of the analysed structure, i.e. the number of equations in the system of Eq. (4.67)), or the system is solved directly.

The general solution of a homogeneous system of differential equations has the form (see also [4.2])

$$\{X_j\} = \sum_{n=1}^{k} C_n \{^nX_j\} \tag{4.69}$$

where ^nX_j are the functions of the fundamental system and C_n are constants. It is assumed that the solution has the form

$$X_j = k_j e^{\lambda\bar{\varphi}} \tag{4.70}$$

---

*) For a homogeneous structure $\varkappa = 1$, hence $[_x\delta_{ij}] = [\delta_{ij}]$; Eq. (4.67) is then transformed into the simple form of Eq. (4.15).

Introducing Eq. (4.70) into the homogeneous system of Eq. (4.67), we obtain a system of homogeneous linear algebraic equations

$$[[\delta_{ij}] \lambda + [_x\delta_{ij}]] \{k_j\} = 0 \tag{4.71}$$

The solution of the system of Eq. (4.71)

$$\text{Det} [[\delta_{ij}] \lambda + [_x\delta_{ij}]] = 0 \tag{4.72}$$

results in an algebraic equation of k-th degree for λ, having k roots λ_1, λ_2, ..., λ_n, ..., λ_k.

The quantity λ_1 is substituted into the system of Eq. (4.71) and the solution of this system yields the vector $\{^1k_j\}$ which is dependent on one arbitrarily chosen parameter. Referring to Eq. (4.70), the first series of functions of the fundamental system, corresponding to the vector $\{^1k_j\}$, is then

$$\{^1X_j\} = e^{\lambda_1 \bar{\varphi}} \{^1k_j\} \tag{4.73}$$

Similarly, we obtain for λ_n

$$\{^nX_j\} = e^{\lambda_n \bar{\varphi}} \{^nk_j\} \tag{4.74}$$

and, in accordance with Eq. (4.69), we can write the general solution of the homogeneous system of differential equations (Eq. (4.67)). The particular solution of the non-homogeneous system may be found by means of the method of variation of constants using the functions of the fundamental system.

The constants $'C_n$ of the general solution of the non-homogeneous system of differential equations (Eq. (4.67)) have to be determined from the known values of the redundants X at the time of the beginning of action of the structure in the investigated static system.

In this way, it is possible to analyse — as long as the applied creep coefficient satisfies the requirement of affinity — all cases of behaviour of a structure, as they were described in Sect. 4.1.1.1 which dealt with the analysis of homogeneous structures (effect of changes of a static system, interventions into the geometry of the structure, and also the stress history of a structure cast on a centering; this stress state, in contrast to a homogeneous structure, is variable in time and does not correspond to the stress state of an elastic structure). If, however, the analysis of the time-dependent distribution of the behaviour of a homogeneous structure could be carried out by formulae directly applicable to most practical instances, the analysis of non-homogeneous structures is much more complex.

4.1.2 Method of time-discretization

In an analysis of the effects of creep and shrinkage the accuracy depends on the choice of the relations describing these phenomena. The classical analysis based on the rate-of-creep theory (Sect. 4.1.1) is very suitable for mathematical operations;

the analytical procedure yields results in the form of simple formulae when a structure is homogeneous, or when the phenomena may be considered to have affinity. In these cases, the analytical approach is reasonable especially in providing a qualitative assessment as to the significance of the rheological phenomena for the given structure, and furnishing grounds for the decision, if it is necessary at all, to consider these phenomena in the particular case.

The rate-of-creep theory, however, is essentially deficient in neglecting other strain components (especially of the delayed elastic strain — see Sect. 2.1.2). For this reason, this theory is unable to express the effect of the history of loading, and its application to the analysis of structures may be the cause of a gross distortion of the results.

This case is analogous, for example, to the analysis of relaxation (Sect. 2.4.2): the analysis based on the rate-of-creep theory was described by Eq. (2.64) whose results, however, were different from the results obtained by the numerical method of time-discretizations using the general creep function.

As in the calculations of strain and stress variations in concrete members (Chapter 2), the analysis of structures can be carried out employing the techniques based on time-discretization.

The analysis aims to investigate the development of the internal forces in a statically indeterminate structure. Due to the restraint of free deformations (or, vice versa, due to the inducement of deformation impulses), statically indeterminate quantities originate and develop with time in the structure which characterize its stress state.

The principle of the method is the division of the time of existence of the statically indeterminate structure in the given static system into a finite number of relatively short partial time intervals. The internal forces and the deformations at the beginning of the investigated interval are known from the previous analysis (or from the initial condition at the commencement of the analysis). The external loading of the structure (together with the statically indeterminate quantities from the preceding time intervals) are assumed to be constant in the course of this interval — the variations of the loading are taken into account at the change of the time intervals. It is assumed that the increments $\{\Delta X\}$ of the statically indeterminate quantities will start to act on the structure at the mid-point of the investigated time interval by a value which is constant until the end of this interval. The magnitude of the increments $\{\Delta X\}$ of the statically indeterminate quantities arises from the condition that these increments must eliminate the increments of the deformations on the primary system at the points of action of the redundants, these deformations having developed during the investigated time interval.

Evidently, the concept of this procedure is similar to the second variant of the analysis of the stress relaxation (see the conclusion of Sect. 2.4, Eqs. (2.88) and (2.90)). The method can be modified to allow for the prescribed deformations at the points of action of the redundants.

182

In the calculation of the increments of deformations, the entire history of the loading process must be respected, i.e. the effects of all the loads, which acted on the structure at some period of its existence, must be considered. The computation is then very tedious, but it can be simplified considerably if the special form of creep function is used (Sect. 2.3.4).

A significant reduction of the analysis can also be achieved by using the rate-of-creep theory*) which does not consider the history of the loading process and thus neglects the effects which do not directly concern the investigated time interval. However, with regard to the inherent defects of the rate-of-creep theory and the ensuing unreliability of its results, the analysis of structures based on the time-discretization method is still unnecessarily complicated.

The method of time-discretization leads to a multitude of relatively simple, but several repeated calculations. Hence, the technique is suited for computers.

4.1.3 Effective time method of action of redundants

Apart from the general method of time-discretization (Sect. 2.4.3) for the analysis of the stress relaxation in concrete members (Sect. 2.4), a simplified technique, termed "effective time method" was demonstrated in Sect. 2.4.4, which yielded results very close to those of the accurate analysis. The concept of the method was the imposition of the full value of the stress decrement at a suitable instant. This instant was determined on the basis of the results obtained from the application of the rate-of-creep theory (Sect. 2.4.3.1).

Hence, an analogy of this technique is possible for the development of a similar method for use in the analysis of structures. First, the structure is analysed by means of the rate-of-creep theory (Sect. 4.1.1), to obtain the time-dependent distributions of the statically indeterminate quantities $X_j(t)$. For example, for a homogeneous structure changing the system of its static behaviour in time t_0, these redundants are generally given by Eq. (4.25)

$$X_j(t) = [X_j(t_0) - {}_eX_j] e^{-[\varphi(t) - \varphi(t_0)]} + {}_eX_j \tag{4.75}$$

In most cases, when the structure has been made monolithic at instant t_0 and hence, when the value of the desired statically indeterminate quantity is zero at instant t_0 (i.e. $X_j(t_0) = 0$), Eq. (4.75) can be written in the form (compare with Eq. (4.34))

$$X_j(t) = {}_eX_j(1 - e^{-[\varphi(t) - \varphi(t_0)]}) \tag{4.76}$$

Actually, the redundants grow continuously during the period $t_0 \ldots t$. Let us now assume another behaviour of the structure: the statically indeterminate quan-

---

*) If the rate-of-creep theory is used, the time-discretization method for a differential time interval takes the form of differential equations.

tities $(\tilde{X}(t))$ do not start acting since the instant t_0, but they are imposed at an instant τ^* immediately with their full (final) values.

If the external loading of the structure does not vary, the magnitude of the increments of deformations at the points of action of the statically indeterminate quantities is

$$\{\tilde{P}_i\} = [\varphi(t, t_e) - \varphi(t_0, t_e)]\{_0P_i\} \tag{4.77}$$

where t_e designates the beginning of action of the external load (for example, de-centering in the original static system, so that $(t_e < t_0)$, and $\{_0P_i\}$ are the deformations on the elastic structure (see Eq. (4.16)).

The statically indeterminate quantities $\tilde{X}(t)$, acting by their full magnitudes from instant τ^*, will induce deformations, which are $(1 + \varphi(t, \tau^*)) -$ fold the corresponding elastic deformations.

These statically indeterminate quantities could be found by solving a system of linear algebraic equations of the form

$$(1 + \varphi(t, \tau^*))[\delta_{ij}]\{\tilde{X}_j(t)\} + ((\varphi(t, t_e) - \varphi(t_0, t_e))\{_0P_i\} = 0 \tag{4.78}$$

yielding

$$\{\tilde{X}_j(t)\} = -\frac{\varphi(t, t_e) - \varphi(t_0, t_e)}{1 + \varphi(t, \tau^*)}[\delta_{ij}]^{-1}\{_0P_i\} \tag{4.79}$$

which, considering Eq. (4.19), can be written

$$\{\tilde{X}_j(t)\} = -\frac{\varphi(t, t_e) - \varphi(t_0, t_e)}{1 + \varphi(t, \tau^*)}\{_eX_j\} = \zeta(t, \tau^*, t_0, t_e)\{_eX_j\} \tag{4.80}$$

This means that the redundants $\{\tilde{X}_j(t)\}$ induced as the consequence of creep, are proportional to the redundants $\{_eX_j\}$ which would develop in the elastic structure. The coefficient ζ depends on the history of the investigated process and also on the applied theory of creep of concrete.

The time τ^*, at which the statically indeterminate quantities started to act, is not known. However, comparative studies have indicated that the rate-of-creep theory, in spite of many justifiable deficiencies, is sufficiently accurate to determine τ^*.

Using the rate-of-creep theory, the coefficient of proportionality is given by

$$\zeta(t, \tau^*, t_0, t_e) = \zeta(t, \tau^*, t_0) = \frac{\varphi(t) - \varphi(t_0)}{1 + \varphi(t) - \varphi(\tau^*)} \tag{4.81}$$

which, typically for the rate-of-creep theory, is independent of the time t_e of introduction of the external loading.

Comparing the results given by Eqs. (4.76) and (4.80), and considering Eq. (4.81), we obtain

$$1 - e^{-[\varphi(t) - \varphi(t_0)]} = \frac{\varphi(t) - \varphi(t_0)}{1 + \varphi(t) - \varphi(\tau^*)} \tag{4.82}$$

and, whence, it follows

$$\varphi(\tau^*) = 1 + \varphi(t) - \frac{\varphi(t) - \varphi(t_0)}{1 - e^{-[\varphi(t) - \varphi(t_0)]}} \tag{4.83}$$

Equations (4.82) and (4.83) are coincident with Eqs. (2.95) and (2.96) which were introduced in Sect. 2.4.3.1 in the analysis of the relaxation of stress by means of time-discretization in a single step. The subsequent procedure is very similar, too.

The values $\varphi(t)$ and $\varphi(t_0)$ are determined for the given instants t and t_0 using the rate-of-creep theory (for example, with the aid of available Tables); substituting into Eq. (4.83), $\varphi(\tau^*)$ is determined and so is the desired instant τ^*.

The instant may also be determined directly. Starting from Mörsch type expression of creep coefficient

$$\tau^* = \left[\ln \frac{\varphi_\infty^2}{\varphi_\infty^2 - \left[1 + \varphi(t) - \frac{\varphi(t) - \varphi(t_0)}{1 - e^{-[\varphi(t) - \varphi(t_0)]}}\right]^2} \right]^2 \tag{4.84}$$

(compare with Eq. (2.97) for the analysis of relaxation).

Now, if instant τ^* is known, *the general form of the creep coefficient* may be used, and the desired statically indeterminate quantities $\{\tilde{X}_j(t)\}$ are calculated in accordance with Eq. (4.80).

An example is given which explains the analysis of a structural system composed of two simple girders of equal age, of equal span L (Fig. 4.3a), carrying a loading q since time $t_e = 28$ days (de-centering). The girders were connected at time $t_0 = 180$ days and they behave as a continuous girder of two spans in the subsequent period of time. The magnitude of the support bending moment $X(t)$, at time $t = 1\ 180$ days, is required (see Fig. 4.3c).

Using the same characteristics of the concrete cross-section as in the example of the analysis of relaxation (Sect. 2.4.3.1), we arrive at $\tau^* = 405$ days. Considering that the support bending moment on an elastic structure is $_eX = -\frac{1}{8}qL^2$, the general theory of creep (Eq. (4.80)) gives the magnitude of the bending moment of the required support at time $t = 1\ 180$ days:

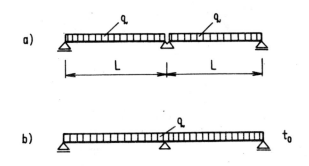

a)

b)

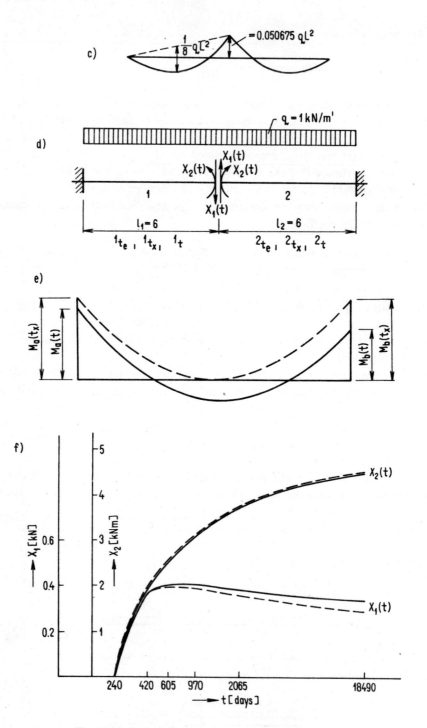

Fig. 4.3 Method of effective time — examples of analysis

$$\tilde{X}(t) = -\frac{\varphi(1\ 180,\ 28) - \varphi(180,\ 28)}{1 + \varphi(1\ 180,\ 405)} \frac{1}{8} qL^2 = -0.405\ 4\ \frac{qL^2}{8} =$$

$$= -0.050\ 675qL^2$$

which is 40.54 % of the value that this moment would attain in a structure already cast as a continuous girder.

If this result is compared with those obtained from the analysis of the same example by means of the relaxation method (see Sect. 4.1.5), which is accurate for the case of a homogeneous structure that has changed its static system, it appears that the difference from the results obtained by a detailed numerical analysis of relaxation is less than 0.5 per cent; however, the results are equal when relaxation is analysed in one step, as demonstrated in Sect. 2.4.3.1.

Let us assume now a structure which consists of n parts (see also [4.5]). Each part has different properties, e.g. age, quality of concrete, environmental conditions, etc. Each part has its own time scale measured from the time of casting. The instant $'t_e$ designates the time when the first load (usually dead load) was applied to the r-th part, $'t_x$ means the time of connecting parts into a new static system, $'t$ is an observed time.

The force method is used for the solution of the problem. The redundants actually begin to act at the instants $'t_x$ and then grow continuously during the intervals $('t_x, 't)$. This continuous process is simply substituted by a constant action of the redundants in intervals $('\tau^*, 't)$. Hence, the redundants begin to act with their full values on each part at the instants τ^*, and are then assumed to be constant until the observed times $'t$.

The displacement of the point i of the structure, developed by the unit statically indeterminate force $X_j('t) = 1$ acting at the point j at the observed time $'t$ (an element of the flexibility matrix $[D^*]$), is given by the formula

$$d_{ij}^* = \sum_{r=1}^{n} {}^r\delta_{ij} \frac{1 + \varphi('t, '\tau^*)}{E_r('\tau^*)}$$

where

$$^r\delta_{ij} = \int_{\Omega_r} \frac{M_i(s)\ M_j(s)}{I(s)}\ ds$$

(Ω_r means integration over the r-th part).

The increase in displacement of the point i resulting from the external load acting since time $'t_e$ on each part of the structure (an element of the right-hand-side vector $\{D_0\}$) is defined as follows

$$d_{i0} = \sum_{r=1}^{n} {}^r\delta_{i0} \frac{\varphi('t, 't_e) - \varphi('t_x, 't_e)}{E_r('t_e)}$$

187

where

$$r\delta_{i0} = \int_{\Omega_r} \frac{M_i(s)\,\mathfrak{M}(s)}{I(s)}\,ds$$

The values of unknown redundants are then found from the system of linear algebraic equations, in accordance with the force method principle:

$$[D^*]\{X\} = \{D_0\}$$

If some deforming impulse is imposed on the structure before its parts are connected into a new static system (e.g. eccentric expansion of an arch in the mid-span in order to affect the distribution of internal forces) it can be considered as a further loading. It is characteristic of this type of loading that the elastic component of deformation is passed before the static system changes.

There are other ways of affecting the static behaviour of the structure, for instance by changing the position of the supports of a continuous girder. In that case, the elastic component of deformation must be involved in the calculation, together with the time-dependent component on the right-hand side of the system of equations.

The effective times $r\tau^*$ depend on the deformation of the whole structure. It is not usually possible to determine the effective times independently of each other.

The following theorem was proved by Bažant [4.3]. If the strain varies linearly with the creep function or with the creep coefficient φ, then the stress varies linearly with the relaxation function. Hence, if the strain is assumed to be varying linearly with φ, the effective time corresponding to each part of the structure may be evaluated separately using the formula (2.96) derived for the relaxation problem, i.e.

$$\varphi_r(r\tau^*) = 1 + \varphi_r(rt) - \frac{\varphi_r(rt) - \varphi_r(rt_x)}{1 - e^{-[\varphi_r(rt) - \varphi_r(rt_x)]}}$$

The accuracy of the presented approach is illustrated by the results of the calculation of a typical structure.

The structure consists of two opposite cantilevers of constant cross-section, each of length 6 m (Fig. 4.3d), carrying the dead load. The ages of the cantilevers differ by about 212 days. The cantilevers were joined when the concrete of the younger cantilever was 28 days old, thus forming a new static system. At the instant when the cantilevers were connected, neither vertical force (the redundant X_1) nor bending moment (the redundant X_2) were acting at the mid-span. The redundants X_1 and X_2 are induced by the creep of concrete, and these significantly affect the distribution of internal forces in the structure. The bending moment diagram, which was symmetrical at the instant of connection (Fig. 4.3e, dashed line), tends to change its shape to that shown by the solid line in Fig. 4.3e.

The analysis is carried out to provide results for several periods (from 6 months to 50 years) covering almost the whole projected lifetime of the structure, and

188

providing information about the development of the internal forces. The effective times are determined as follows:

^1t (days)	$^1\tau^*$ (days)	^2t (days)	$^2\tau^*$ (days)
420	312.9	208	68.9
605	366.3	393	87.9
970	439.9	758	102.2
2 065	560.3	1 853	137.7
18 490	787.7	18 278	181.3

Fig. 4.3f shows the variation of the redundants during the observed period and their deviation from the exact computer solution obtained by the time-discretization method. The mid-span shear force, where lower accuracy is reached, does not affect the final shear forces and bending moment diagrams significantly. The values of moments at the fixed ends predicted by the effective-time method, and their errors in comparison with the exact solution are:

^1t (days)	$M_a(^1t)$ (kNm)	error (%)	$M_b(^1t)$ (kNm)	error (%)
420	− 18.29	0.15	− 13.84	0.20
605	− 17.89	0.72	− 13.02	−0.52
970	− 17.33	0.75	− 12.43	−0.56
2 065	− 16.57	1.26	− 11.91	−1.16
18 490	− 15.54	2.04	− 11.46	−2.04

It can be seen that the agreement between the exact values and those given by the effective-time method is extremely good.

The effective-time method is instructive and versatile and may thus conveniently be used as a design tool. The possibility of easily taking into account the effect of the time of load application illustrates the superiority of the method over the great majority of other approximate methods. The results can be obtained directly from simple hand calculations and the ease of application of the method makes it suitable as an aid for the repeated analyses required during the initial design stages.

4.1.4 Trost–Bažant method for the analysis of statically indeterminate structures

This method (summarized, e.g., in [4.3]) is applicable for structures in which the support conditions change suddenly or change at the same rate as creep.

Basically the force method is used and the flexibility coefficients have to be evaluated in different ways for the action of:

(i) sustained loads (statically determinate forces and known redundants) and
(ii) unknown statically indeterminate time-dependent forces.

In the first case, (i), the time-dependent displacements equal the elastic displacement multiplied by $\varphi(t, t_0)$ (the creep coefficient at time t for a load applied at age t_0). In the second case, (ii), the time-dependent displacements due to time-dependent redundants (which are zero at age t_0) have to be evaluated by multiplying the elastic displacements (the flexibility coefficients) by the terms

$$1 + \chi\varphi(t, t_0)$$

where χ is the aging coefficient (see Chapter 2), which is necessary because the redundants develop gradually with time.

4.1.5   Relaxation method

A simple method intended for practical use as a design tool has been developed for the analysis of the effects of creep and shrinkage upon the response of concrete structures; it is very effective especially in the cases of complex structures which change their system of static behaviour in the course of the construction process. In this method, the duration of the structure is subdivided into a few time intervals. In each of those intervals, the analysis is carried out in two consecutive steps. In the first step, the relaxation is analysed (in accordance with an appropriate creep theory, and therefore not just the rate-of-creep theory), and, in the second step, the routine procedures developed for the analysis of elastic structures are utilized. The method has a very simple algorithm, it requires only very little specialized knowledge, and an arbitrary accuracy of the results may be reached by varying the fineness of the partial time intervals. The method is especially convenient for structures, for which elastic analysis by computer programs are available; at present, this applies to frame structures (plane as well as space frames, including arches and grillages), while another advantage is the possibility of a repeated utilization of influence lines which has been developed for bridge structures at the earlier stage for the evaluation of the effect of live load.

The principle of the method is demonstrated by the following analysis in a typical time interval.

Let the state of stress and the deformations of a structure be known at time t_0. The internal forces at this instant are described by the vector $\{\mathfrak{S}\}$. The state of stress and the deformations of the structure at time $t > t_0$ are to be determined.

It is typical that the constant internal forces $\{\mathfrak{S}\}$ (the axial force, bending and torsional moments and shear forces) were applied suddenly at age t_1 on the structure in its initial structural system which may be statically determinate or indeterminate. At time t_0 the structural system can be changed e.g. by adding further constraints without any sudden change of stresses at time t_0.

190

First step of the analysis

Let us assume that at instant t_0, when the structure is stressed by the internal forces $\{\mathfrak{S}\}$, it is restrained in its entire extent in order to prevent any further deformation changes (for $t > t_0$), i.e., deformation of the structure remains the same as at time t_0.

When the deformation of the entire structure remains constant, the strain of each of its elements must also be constant; hence, it must hold

$$\frac{d}{dt}\{\varepsilon(t)\} = 0 \tag{4.85}$$

In this way (in accordance with the theory used for the prediction of creep), we now analyse the stress relaxation (see Sect. 2.4), i.e. a phenomenon in which the changes of strain are prevented and the internal forces decrease from their original value $\{\mathfrak{S}\}$ at the beginning of the investigated time interval to the value $\{\mathfrak{S}_1(t)\}$ at the investigated time t.

It follows from Eq. (4.85) that

$$\{\Delta\varepsilon(t)\} = \{\varepsilon(t)\} - \{\varepsilon(t_0)\} = 0 \tag{4.85a}$$

Considering the typical stress history

$$\{\mathfrak{S}\} \text{ within the period } t_1 \ldots t_0$$
$$\{\mathfrak{S}_1(t_0)\} = \{\mathfrak{S}\}$$
$$\{\mathfrak{S}_1(t)\} \text{ within the period } t_0 \ldots t$$

and by substituting into Eq. (2.22), the following relation is found

$$\{\varepsilon(t)\} = (1 + \varphi(t, t_1))\,[H(t_1)]\,\{\mathfrak{S}\} +$$
$$+ \int_{t_1}^{t_0} (1 + \varphi(t, \tau))\,[H(\tau)]\,\frac{d}{d\tau}\{\mathfrak{S}\}\,d\tau +$$
$$+ \int_{t_0}^{t} (1 + \varphi(t, \tau))\,[H(\tau)]\,\frac{d}{d\tau}\{\mathfrak{S}_1(\tau)\}\,d\tau$$

Note that the internal forces $\{\mathfrak{S}\}$ are constant during the period $t_1 \ldots t_0$. Thus

$$\frac{d}{d\tau}\{\mathfrak{S}\} = 0$$

and the above relationship will therefore be of the form

$$\{\varepsilon(t)\} = (1 + \varphi(t, t_1))\,[H(t_1)]\,\{\mathfrak{S}\} +$$
$$+ \int_{t_0}^{t} (1 + \varphi(t, \tau))\,[H(\tau)]\,\frac{d}{d\tau}\{\mathfrak{S}_1(\tau)\}\,d\tau$$

Similarly,

$$\{\varepsilon(t_0)\} = (1 + \varphi(t_0, t_1)) [H(t_1)] \{\mathfrak{S}\}$$

The condition that the deformations are frozen from time t_0, given by Eq. (4.85a), causes stress variation $\{\mathfrak{S}_1(t)\}$ which satisfies at every location s the equation

$$\{\Delta\varepsilon(t)\} = \int_{t_0}^{t} (1 + \varphi(t, \tau)) [H(\tau)] \frac{d}{d\tau} \{\mathfrak{S}_1(\tau)\} \, d\tau +$$
$$+ (\varphi(t, t_1) - \varphi(t_0, t_1)) [H(t_1)] \{\mathfrak{S}\} = 0 \qquad (4.85b)$$

Eq. (4.85b) represents a Volterra integral equation for function $\mathfrak{S}_1$, which may be easily solved numerically with high accuracy in a step-by-step manner (see Sections 2.4.3.1 and 2.4.3.3). For the simplified solution the effective time method (see Section 2.4.3.2) or the age-adjusted effective modulus method (Section 2.4.3.7) can be applied.

In the assumed restraint, which applies to the whole structure continuously, reactions develop in the course of the relaxation process; these reactions act on the structure by a load $- \{q_2(t)\}$ at time t. The components of this load are

$$\{q_2\}^T = q_{2x}, q_{2z}, q_{2y}, m_{2t}, m_{2y}, m_{2z} \qquad (4.86)$$

where $-q_{2x}$, $-q_{2z}$ and $-q_{2y}$ designate continuous loading by the reactions in the longitudinal direction, in the directions perpendicular to the y-axis and to the z-axis, respectively; $-m_{2t}$, $-m_{2y}$ and $-m_{2z}$ designate loading by the reaction moments rotating about the axes x, y and z, respectively.

The reactions $- \{q_2(s, t)\}$ are determined on the basis of the change of the internal forces developing in the structure in a state of constant deformation during the period $t - t_0$, i.e. on the basis of the difference $\{\mathfrak{S}\} - \{\mathfrak{S}_1(t)\}$.

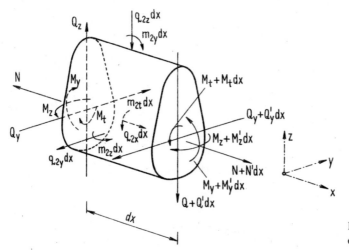

Fig. 4.4 Equilibrium of the element of a straight member

If the diagrams of the internal forces $\{\mathfrak{S}(s)\} - \{\mathfrak{S}_1(s, t)\}$ are described by continuous functions, then

$$\{q_2(s, t)\} = [d](\{\mathfrak{S}(s)\} - \{\mathfrak{S}_1(s, t)\}) = [d]\{\mathfrak{S}_2(s, t)\} \tag{4.87}$$

where

$$\{\mathfrak{S}_2(s, t)\} = \{\mathfrak{S}(s)\} - \{\mathfrak{S}_1(s, t)\} \tag{4.88}$$

and where $[d]$ is a matrix of operators whose form for the case of a straight member*) is

$$[d] = \begin{bmatrix} -\dfrac{\partial}{\partial x} & 0 & 0 & 0 & 0 & 0 \\[2mm] & 0 & 0 & 0 & -\dfrac{\partial}{\partial x} & 0 \\[2mm] & & 0 & 0 & 0 & -\dfrac{\partial}{\partial x} \\[2mm] \text{symmetr} & & -\dfrac{\partial}{\partial x} & 0 & 0 \\[2mm] & & & & 1 & 0 \\[2mm] & & & & & -1 \end{bmatrix} \tag{4.89}$$

At the points where the functions $\{\mathfrak{S}(s)\} - \{\mathfrak{S}_1(s, t)\}$ have a discontinuity, the reactions $\{q_2\}$ are expressed by point loads N_2, Q_{z2} and Q_{y2} and by the moments M_{y2}, M_{z2} and M_{t2}. Hence, the result of the analysis in the first step is the evaluation of the internal forces $\mathfrak{S}_1$, which develop at the individual cross-sections s during the period $t_0 \dots t$ in which the deformation of the structure has been kept constant and also the determination of the load q_2.

This data enables the analysis to continue to the second step.

Second step of the analysis

The restraint of the structure, assumed in the first step, actually does not exist; consequently, the actual structure is now to be subjected to a loading by the inverted reactions originating at time $t > t_0$, i.e. by loadings $\{q_2(s, t)\}$ acting as an external loading. By analysing the structure loaded in this way, the internal forces $\{S_2(s, t)\}$ are obtained.

The resulting stress state of the structure at time t is characterized by the internal forces $\{S(s, t)\}$, which are obtained as a sum of the internal forces of both the steps of the analysis, viz.

$$\{S(s, t)\} = \{\mathfrak{S}_1(s, t)\} + \{S_2(s, t)\} \tag{4.90}$$

---

*) The coordinate measured along the median line of the straight member is designated as x (Fig. 4.4).

The second step of the analysis can be conveniently carried out by assuming the structure to be elastic. This is much easier than with the use of other methods (resulting, e.g. in sets of differential equations), because computer programs for the analysis of statically indeterminate structures are currently available, or influence lines, developed for the evaluation of the effect of live load, may be again exploited. Thus, the effect of creep analysed in this way may be included among the other loading cases and the entire analysis of all the loading cases may be carried out simultaneously.

The practical analysis using the principles of the relaxation method is demonstrated on the following simple example. Since a plane girder structure is considered, internal forces are solely characterized by the bending moments $\mathfrak{M}$ and the effect of shear forces may be neglected.

Two simple girders of a span L and carrying a load q (Fig. 4.5a) which were de-centred (i.e. subjected to a load for the first time) at the age $t_e = 28$ days, were connected over their common support at the age $t_0 = 180$ days (Fig. 4.5b). For any subsequent period of time, they behave as a continuous beam of two spans. The stress state of the structure at time $t = 1\,180$ days is to be determined.

Both girders are of the same age, and hence the structure is homogeneous. The cross-sectional area of the girder is 1.08×10^6 mm^2, the cross-sectional perimeter is $4\,800$ mm, and the structure is stored in free space.

The diagram of the bending moments $\mathfrak{M}$ at the instant when the two girders were connected is shown in Fig. 4.5c. Within the period $t_0 \ldots t$, assuming a constant deformation (first step of the method, Eq. (4.85)), the relaxation process takes place. By finding the relaxation (see the example in Sect. 2.4.3.1), we determine the value of the stress of the concrete member with the above characteristics which was subjected to the load at the age of 28 days, and, since its deformation remains constant between $t_0 = 180$ days and $t = 1\,180$ days, the stress decreases to $0.592\,7$ of its original value in the course of the relaxation process (Table 2.1).

This means that, assuming a constant deformation of the structure (the first step of the method), the original bending moments $\mathfrak{M}$ (Fig. 4.5c) decrease to

$$\mathfrak{M}_1(x, t) = 0.592\,7\mathfrak{M}(x) \tag{4.91}$$

Their diagram is shown in Fig. 4.5d.

The moments $\mathfrak{M}_2(x, t)$, according to Eq. (4.88), are

$$\mathfrak{M}_2(x, t) = \mathfrak{M}(x) - \mathfrak{M}_1(x, t) = 0.407\,3\mathfrak{M}(x) \tag{4.92}$$

and they are portrayed in Fig. 4.5e.

The loading for the second step of the analysis (Fig. 4.5f) is

$$q_2(x, t) = -\frac{\partial^2 \mathfrak{M}_2(x, t)}{\partial x^2} = 0.407\,3\,q \tag{4.93}$$

The structure in its new static arrangement (a continuous beam of two spans)

194

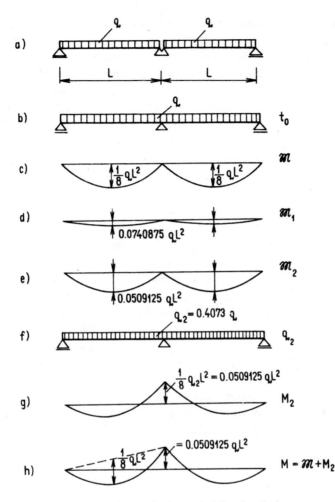

Fig. 4.5 Relaxation method — principle of analysis

is now, as an elastic structure, subjected to this load q_2 (Fig. 4.5f), and the developed bending moments $M_2(x, t)$ are portrayed in Fig. 4.5g.

The desired diagram of bending moments on the structure at time t is easily determined from Eq. (4.90), i.e. by summing the diagrams of Figs. 4.5d and 4.5g. The diagram of the resulting bending moments is shown in Fig. 4.5h.

The analysis is accurate for a homogeneous structure, and is independent of the length of the time interval Δt.

For non-homogeneous structures, the analysis becomes more accurate as the non-homogeneity of the structure decreases and as the number of the time intervals increases. The accuracy of the analysis also depends on the condition of whether the investigated effect would manifest itself in a homogeneous structure, or not.

This should be appreciated when chosing the number of time intervals. For structures having the age of concrete varying in succession (segmental cantilevering bridges with several casting carriages, casting on sliding centering, slipped-out bridges, etc.), only a few intervals (often a single one for each of the successive static systems) are usually necessary for a reasonably accurate expression of the actual stress. Evidently, the number of intervals Δt (for a homogeneous and non-homogeneous structure) for a structure changing its system several times must be chosen so that at least one time interval Δt corresponds to each structural system (i.e. a change of the structural system must always be the start of the time intervals).

The described relaxation method is very convenient for the analysis of external interventions into the geometry of a structure. If the magnitude of these interventions does not vary in the course of the investigated time interval, the solution for a homogeneous structure is given by the first step of the method where a constant deformation is maintained for the entire time interval. The reactions $\{q_2\}$, which are merely manifested as the forces or moments in the direction of the external interventions, together with the effects causing the interventions at the beginning of the time interval, will ensure the equilibrium at the points of the reactions.

In a non-homogeneous structure, a continuously distributed loading due to the reactions $\{q_2\}$ develops, but its effect in a structure, which is not extremely non-homogeneous, is small as compared with the results of the first step of the analysis.

The above given procedure allows the determination of the stress state in a structure at time t. Another necessary task is to determine the deformation of the structure. Bearing in mind that the stress state of the structure, characterized in typical cases by internal forces

$$\text{within the period } t_1 \ldots t_0 \qquad \{S(t)\} = \{\mathfrak{S}\}$$
$$\text{and within the period } t_0 \ldots t \quad \{S(t)\} = \{\mathfrak{S}_1(t)\} + \{S_2(t)\},$$

(see Eq. (4.90)), is already known, we can determine the time variation of the deformation increments after time t_0 of all members of the structure by substituting the internal forces into Eq. (2.22)

$$\{\Delta\varepsilon(t)\} = \{\varepsilon(t)\} - \{\varepsilon(t_0)\} =$$
$$= (1 + \varphi(t, t_1)) [H(t_1)] \{S(t_1)\} +$$
$$+ \int_{t_1}^{t} (1 + \varphi(t, \tau)) [H(\tau)] \frac{d}{d\tau} \{S(\tau)\} \, d\tau -$$
$$- (1 + \varphi(t_0, t_1)) [H(t_1)] \{S(t_1)\} -$$
$$- \int_{t_1}^{t_0} (1 + \varphi(t, \tau)) [H(\tau)] \frac{d}{d\tau} \{S(\tau)\} \, d\tau =$$
$$= (1 + \varphi(t, t_1)) [H(t_1)] \{\mathfrak{S}\} +$$

$$+ \int_{t_1}^{t_0} (1 + \varphi(t, \tau)) [H(\tau)] \frac{\mathrm{d}}{\mathrm{d}\tau} \{\mathfrak{S}\} \, \mathrm{d}\tau +$$

$$+ \int_{t_0}^{t} (1 + \varphi(t, \tau)) [H(\tau)] \frac{\mathrm{d}}{\mathrm{d}\tau} (\{\mathfrak{S}_1(\tau)\} + \{S_2(\tau)\}) \, \mathrm{d}\tau -$$

$$- (1 + \varphi(t_0, t_1)) [H(t_1)] \{\mathfrak{S}\} -$$

$$- \int_{t_1}^{t_0} (1 + \varphi(t, \tau)) [H(\tau)] \frac{\mathrm{d}}{\mathrm{d}\tau} \{\mathfrak{S}\} \mathrm{d}\tau$$

Since it holds that $\dfrac{\mathrm{d}}{\mathrm{d}\tau} \{\mathfrak{S}\} = 0$, it is possible to write

$$\{\Delta\varepsilon(t)\} = \int_{t_0}^{t} (1 + \varphi(t, \tau)) [H(\tau)] \frac{\mathrm{d}}{\mathrm{d}\tau} \{\mathfrak{S}_1(\tau)\} \, \mathrm{d}\tau +$$

$$+ (\varphi(t, t_1) - \varphi(t_0, t_1)) [H(t_1)] \{\mathfrak{S}\} +$$

$$+ \int_{t_0}^{t} (1 + \varphi(t, \tau)) [H(\tau)] \frac{\mathrm{d}}{\mathrm{d}\tau} \{S_2(\tau)\} \, \mathrm{d}\tau$$

The first two terms of this relationship represent the relaxation condition (4.85b) expressing that the deformations are frozen in the first step of analysis. Thus, the final expression for the deformation increments is

$$\{\Delta\varepsilon(t)\} = \int_{t_0}^{t} (1 + \varphi(t, \tau)) [H(\tau)] \frac{\mathrm{d}}{\mathrm{d}\tau} \{S_2(\tau)\} \, \mathrm{d}\tau \tag{4.94}$$

Thus only the component of internal forces $\{S_2\}$ produces deflection changes after t_0. The component $\{\mathfrak{S}_1\}$, corresponding to the stress variation in the first step of analysis during which deformations are imagined to be frozen, has no influence upon the deflection changes.

The deformation of the structure can then be determined (after integration in time) by applying the usual methods of statics for beams and frame structures (equations of deflection lines, Mohr's theorem, principle of virtual works, etc).

For fine time intervals (i.e. for a short time period $t_0 \ldots t$) it may be approximately written that

$$1 + \varphi(t, \tau) \simeq 1 \tag{4.95}$$

The approximate character of this relationship consists in neglecting the creep deformations caused by the internal forces $\{S_2\}$ during the investigated interval.

Considering also the modulus of elasticity constant in time, Eq. (4.94) takes a very simple form

$$\{\Delta\varepsilon(t)\} = [H] \{S_2(t)\} \tag{4.96}$$

which corresponds to the deformation of members in an elastic structure subjected to the loading $\{q_2\}$.

197

This approach allows a substantial simplification of the analysis of deformations, because common methods of analysing elastic structures are utilized; consequently, the deformations can be calculated simultaneously by the analysis of the relaxation method which transforms the analysis in the second step to that of an elastic structure.

However, Eq. (4.96) always yields only approximate results for non-homogeneous, and also for homogeneous structures. The accuracy of the analysis is improved when the time intervals are shortened so that the introduced simplifications become justified.

No accurate results are obtained even for statically determinate structures. For example, in a simple beam of a span L and flexural rigidity EI, subjected to a uniformly distributed load q constant in time (Fig. 4.6), the value of the deflection at mid-span, caused by creep within the time interval $\langle t_0, t_k \rangle$ is, according to the rate-of creep theory,

$$w = \frac{5}{384} \frac{qL^4}{EI} [\varphi(t_k) - \varphi(t_0)] = \frac{5}{384} \frac{qL^4}{EI} \Delta\varphi \tag{4.98a}$$

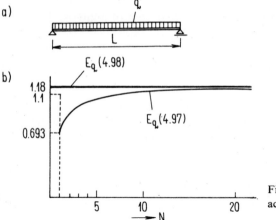

Fig. 4.6 Convergency of deflection to the accurate value in relation to the number of time intervals

If the above approximate method is adopted (in accordance with Eq. (4.96)), this deflection is expressed by the relation

$$w = \frac{5}{384} \frac{qL^4}{EI} \sum_{n=1}^{N} (1 - e^{-\Delta\varphi_n}) \tag{4.97}$$

where N is the number of intervals into which the period $t_k \to t_0$ has been subdivided; n designates the n-th of these intervals.

If we study, for example, the deflection caused by creep from the age of concrete $t_0 = 1$ month until $t_k = 1$ year, the convergence of Eq. (4.97) to Eq. (4.98a) depends on the number of partial intervals N, as can be seen in Table 4.1 and Fig. 4.6. Evidently, the calculation of the deformation of a structure is much

Table 4.1

N	1	2	4	6	11	22	Equation
$\dfrac{384EI}{5qL^4}\,w$	0.693	0.891	1.022	1.073	1.116	1.148	(4.97)
				1.18 accurate value according to Eqs. (4.98) and (4.98a)			

more sensitive to the length of the time interval, than that of the evaluation of the internal forces in a structure by means of the relaxation method.

In the case of a homogeneous structure it holds that

$$\{S_2(t)\} = \{_eS\}\left(1 - \frac{\mathfrak{S}_1(t)}{\mathfrak{S}}\right)$$

in which $\{_eS\}$ are internal forces of *the elastic structure in its final structural system caused by full external loading.*

Eq. (4.94) thus can be written in the form

$$\{\Delta\varepsilon(t)\} = \int_{t_0}^{t}(1 + \varphi(t, \tau))\,[H(\tau)]\,\frac{d}{d\tau}\left(\{_eS\} - \left\{_eS\,\frac{\mathfrak{S}_1(\tau)}{\mathfrak{S}}\right\}\right)d\tau =$$

$$= -\left\{\frac{_eS}{\mathfrak{S}}\right\}\int_{t_0}^{t}(1 + \varphi(t, \tau))\,[H(\tau)]\,\frac{d}{d\tau}\{\mathfrak{S}_1(\tau)\}\,d\tau$$

which, with reference to Eq. (4.85b), yields

$$\{\Delta\varepsilon(t)\} = (\varphi(t, t_1) - \varphi(t_0, t_1))\,[H(t_1)]\,\{_eS\}$$

Using the notation

$$\{_e\varepsilon\} = [_eH]\,\{_eS\}$$

in which $\{_e\varepsilon\}$ are deformations of the elastic structure, whose flexibility characteristics are given by the matrix $[_eH]$, due to internal forces $\{_eS\}$.

Bearing in mind the diagonal character of matrices $[_eH]$ and $[H(t_1)]$, it is possible to write

$$\{\Delta\varepsilon(t)\} = (\varphi(t, t_1) - \varphi(t_0, t_1))\,\frac{E}{E(t_1)}\,\{_e\varepsilon\}$$

Since the ratio of moduli $E/E(t_1)$ is the same at every location of the homogeneous structure, the deflection increase is expressed in the form

$$\Delta w(s, t) = {}_ew(s)\,(\varphi(t, t_1) - \varphi(t_0, t_1))\,\frac{E}{E(t_1)} \tag{4.98}$$

Thus it can be concluded that *the increment of deflections* of a homogeneous structure is *exactly equal to the deflections of an elastic structure in its final* structural system caused by full applied load, *multiplied by the increase in the creep coefficient value corrected by the ratio of yield moduli.*

Note that E in Eq. (4.98) is that modulus of elasticity which has been used for evaluation of the elastic deflection $_ew$. It need not be the actual Young's modulus of concrete, but any convenient value, e.g. the unity. In any case the value of the product E_ew remains the same and thus the results of formula (4.98) are not influenced by the value of E used in the elastic analysis.

This statement also may be used when estimating the increment of deflections of a non-homogeneous structure whose creep properties (namely the age of concrete) are not too different in individual parts of the structure.

It should be noted that the calculation in the first step of the relaxation method is really the analysis of the stress relaxation, and it is not related, in principle, to any definite theory of creep.

It is generally known that the rate-of-creep theory does not express the actual creep behaviour correctly. In the subsequent discussion, this theory is used only to illustrate the analysis in the first step of the method, and because it leads to simple relations and to solutions in closed form.

In this case, the relaxation condition (Eq. (4.85)), with consideration of Eq. (2.65) has the form of a differential equation

$$[H]\left(\frac{d}{dt}\{\mathfrak{S}_1(t)\} + \{\mathfrak{S}_1(t)\}\frac{d\varphi(t)}{dt}\right) + \{\omega\}\frac{d\varphi(t)}{dt} + \frac{d}{dt}\{\varepsilon_T(t)\} = 0 \quad (4.99)$$

The general solution of Eq. (4.99) may be written in the form

$$\{\mathfrak{S}_1(t)\} = \{C\}\,e^{-\varphi(t)} - [H]^{-1}(\{\omega\} + e^{-\varphi(t)}\{I(t)\}) \quad (4.100)$$

where C are constants
and where

$$\{I(t)\} = \int \frac{d}{dt}\{\varepsilon_T(t)\}\,e^{\varphi(t)}\,dt$$

Equation (4.100) may be further arranged into

$$\{\mathfrak{S}_1(t)\} = \{C\}\,e^{-\varphi(t)} - [K](\{\omega\} + e^{-\varphi(t)}\{I(t)\}) \quad (4.101)$$

where $[K]$ is the diagonal matrix of the constants of stiffness of the cross-section, i.e.

$$[K] = [H]^{-1} = \begin{bmatrix} EA, & 0, & 0, & 0, & 0, & 0 \\ & EI_y, & 0, & 0, & 0, & 0 \\ & & EI_z, & 0, & 0, & 0 \\ & & & GI_t, & 0, & 0 \\ & & & & \dfrac{GA}{\gamma_z}, & 0 \\ \text{symmetr} & & & & & \dfrac{GA}{\gamma_y} \end{bmatrix}$$

200

At time t_0, this structural member was stressed by the internal forces $\{\mathfrak{S}(s)\}$. To determine the constants $\{C\}$ in the general solution of Eqs. (4.100) or (4.101), we have the initial condition

$$\{\mathfrak{S}_1(t_0)\} = \{\mathfrak{S}\}$$

and hence

$$\{C\} = [\{\mathfrak{S}\} + [K](\{\omega\} + e^{-\varphi(t_0)}\{I(t_0)\})]\, e^{\varphi(t_0)}$$

The required internal forces $\{\mathfrak{S}_1(t)\}$, acting at cross-section s, which are necessary for the maintenance of the deformations, are

$$\begin{aligned}\{\mathfrak{S}_1(t)\} = &\;(\{\mathfrak{S}\} + [K]\{\omega\})\, e^{-\Delta\varphi_{t,t0}} - \\ &- [K][\{\omega\} + (\{I(t)\} - \{I(t_0)\})\, e^{-\varphi(t)}]\end{aligned} \qquad (4.102)$$

where

$$\Delta\varphi_{t,t_0} = \varphi(t) - \varphi(t_0) \qquad (4.103)$$

The described procedure for the analysis of the effect of creep in non-homogeneous structures has been introduced in a general form, with consideration of all internal forces at the cross-sections of members in a three-dimensional structure; it also includes the effects of shrinkage and of the variations of temperature. Frequently, when a structure can be modelled by a plane frame, when the effects of shrinkage and of the variations of temperature may be neglected, or when these effects can be analysed independently of the flexural effects, and when it is possible to separate the effect of normal forces (a simple instance of such a structure may be a continuous beam, frequently employed in the construction of bridges), only the distribution of the bending moments is important for the stress of the structure. The analysis by the described technique is then considerably simplified. This technique, using the rate-of-creep theory, will be demonstrated for the sake of simplicity and illustration.

At the beginning of the investigated time interval Δt, i.e. at time t_0, the structure is stressed by bending moments $\mathfrak{M}(x)$ whose distribution at this instant is known. The vector of the internal forces $\{\mathfrak{S}(s)\}$ is then reduced to

$$\{\mathfrak{S}(s)\} = \mathfrak{M}(x)$$

The internal forces $\{\mathfrak{S}_1(s, t)\}$, which are necessary for the maintenance of a constant deformation of the structure within the time interval Δt and are described by the general relation (Eq. (4.102)), are

$$\{\mathfrak{S}_1(s, t)\} = \mathfrak{M}_1(x, t) = \mathfrak{M}(x)\, e^{-\Delta\varphi_{t,t0}} - \qquad (4.104)$$

Furthermore

$$\{\mathfrak{S}_2(s, t)\} = \mathfrak{M}_2(x, t) = \mathfrak{M}(x) - \mathfrak{M}_1(x, t) = \mathfrak{M}(x)\,(1 - e^{-\Delta\varphi_{t,t0}}) \qquad (4.105)$$

201

and the loading $\{q_2\}$ for the second step of the analysis (if the function: $\mathfrak{M}(x) - \mathfrak{M}_1(x, t)$ as well as its first derivative are continuous), is

$$\{q_2(s, t)\} = q_2(x, t) = -\frac{\partial^2 \mathfrak{M}_2(x, t)}{\partial x^2} \tag{4.106}$$

At the cross-section, where the function $\mathfrak{M}(x) - \mathfrak{M}_1(x, t)$ has not a continuous derivative, a concentrated force pertains to the loading q_2, and where the function is not continuous, a corresponding moment also appears.

In the second step of the analysis, the elastic structure is subjected to this loading q_2, and bending moments $M_2(x, t)$ are obtained which correspond to the internal forces $\{S_2(s, t)\}$ in the general formulation of the method. The sought resulting stress of the structure at time t is expressed by Eq. (4.90), which, in this case, has the form

$$M(x, t) = \mathfrak{M}_1(x, t) + M_2(x, t) \tag{4.107}$$

The convergence of the described method to the direct analysis (Sect. 4.1.1.2) will be demonstrated on a simple example.

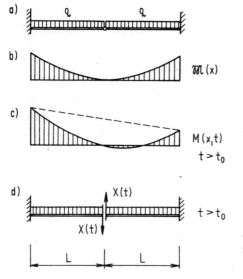

Fig. 4.7 Opposite cantilevers of different age of concrete. Changes of distribution of the bending moments due to creep; shear force developing in the middle hinge

Two opposite concrete cantilevers of constant cross-section, each of length L, carrying a load q (Fig. 4.7a) differ in age. While the left-hand cantilever, which is very old, is practically uninfluenced by creep, creep is very manifest in the right-hand cantilever (young concrete). At the age of the younger (right-hand) cantilever (t_0), both cantilevers have been connected by a hinge. The problem is to find the distribution of the bending moments on the structure at a general time $t > t_0$.

202

At the instant when the cantilevers were connected, no vertical force was acting at the connecting hinge. Due to creep, the right-hand (younger) cantilever deflects more than the older cantilever; consequently, a force $X(t)$ in the hinge develops with time (Fig. 4.7d). The bending moment diagram, which was symmetrical at the instant t_0 of connection (Fig. 4.7b), changes its shape to that pictured in Fig. 4.7c due to the action of the force $X(t)$.

For the demonstration of the analysis by means of the relaxation method, the entire period of behaviour of the connected structure is considered as a single time interval Δt. The bending moments, having a symmetrical diagram at time t_0 (see Figs. 4.7b and 4.8a), must decrease to the values $\mathfrak{M}_1(x, t)$ in order to maintain deflection line at the age of concrete t of the younger cantilever. The values $\mathfrak{M}_1(x, t)$ are plotted by dashed lines in Fig. 4.8a.

a)

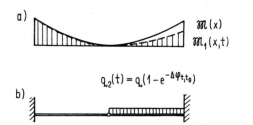

$\mathfrak{M}(x)$
$\mathfrak{M}_1(x, t)$

$q_2(t) = q(1 - e^{-\Delta \varphi t; t_0})$

b)

Fig. 4.8 (a) first step of analysis; (b) second step of analysis

It follows that

$$\mathfrak{M}_1(x, t) = \mathfrak{M}(x)\, e^{-\Delta \varphi t, t_0} \tag{4.108}$$

Furthermore

$$\mathfrak{M}_2(x, t) = \mathfrak{M}(x) - \mathfrak{M}_1(x, t) = \mathfrak{M}(x)\, (1 - e^{-\Delta \varphi t, t_0}) \tag{4.109}$$

In this case, the loading

$$q_2(t) = q(1 - e^{-\Delta \varphi t, t_0}) \tag{4.110}$$

corresponds to those moments $\mathfrak{M}_2(x, t)$ (Fig. 4.8b). This loading is now made to act on an elastic structure and it induces, at the age t of the younger cantilever, a shear force in the connecting hinge,

$$X(t) = \frac{3}{16}\, qL(1 - e^{-\Delta \varphi t, t_0}) \tag{4.111}$$

Since the force X characterizes the redistribution of the internal forces resulting from the effect of creep on this non-homogeneous structure, the analysis in this single time interval is completed.

If the time interval $t - t_0$ is subdivided into N parts so that the value $\Delta \varphi / N$ is the same for all the N parts (Fig. 4.9) and if the analysis is carried out N-times,

203

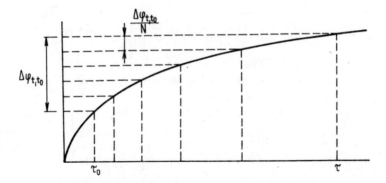

Fig. 4.9 Division of a time interval into parts with equal increments of the creep function

we obtain the following formula for the shear force acting in the connecting hinge

$$X(t) = \frac{3}{16} qL \left[2 - \frac{1}{2^{N-1}} (1 + e^{-\frac{\Delta \varphi_{t,t0}}{N}})^N \right] \tag{4.112}$$

Note:

for $N = 1$, Eq. (4.112) acquires the form

$$X(t) = \frac{3}{16} qL \left[2 - (1 + e^{-\Delta \varphi_{t,t0}}) \right] = \frac{3}{16} qL(1 - e^{-\Delta \varphi_{t,t0}})$$

which is identical with Eq. (4.111).

If the density of division of the time interval increases without limits, i.e. $N \to \infty$, the value of the shear force in the hinge is

$$X(t) = \lim_{N \to \infty} \left\{ \frac{3}{16} qL \left[2 - \frac{1}{2^{N-1}} (1 + e^{-\frac{\Delta \varphi_{t,t0}}{N}})^N \right] \right\} =$$

$$= \frac{3}{8} qL \left\{ 1 - \lim_{N \to \infty} \left[\frac{1}{2^N} (1 + e^{-\frac{\Delta \varphi_{t,t0}}{N}})^N \right] \right\} =$$

$$= \frac{3}{8} qL (1 - \sqrt{e^{-\Delta \varphi_{t,t0}}})$$

which is a result corresponding to the direct analysis (Sect. 4.1.1).

4.2 Analysis of thin-walled and three-dimensional structures

In accordance with the principles presented in the foregoing Section for frame structures, the effects of creep and shrinkage of concrete will now be studied in thin-walled three-dimensional structures.

204

It is again possible to use any theory of creep, but, from the practical point of view, preference is given to methods which allow the calculation of relaxation with a minimum effort.

In the first step of the method, constant deformation of a structure is ensured during the considered time interval. Hence,

$$\frac{\partial}{\partial t} \{\varepsilon^*\} = 0 \tag{4.113}$$

where $\{\varepsilon^*\}$ is the vector of strains defined as follows:

$$\{\varepsilon^*\}^T = \varepsilon_x, \varepsilon_y, \varepsilon_z, \gamma_{xy}, \gamma_{xz}, \gamma_{yz} \tag{4.114}$$

Analysing the relaxation in accordance with Eq. (4.113), we obtain the expression for the stress $\{\sigma_1(t)\}$ at the end of the investigated time interval t_0, t, which is dependent on the stresses $\{\sigma_0\}$ acting on the structure at time t_0 and on the effects of shrinkage and variations of temperature.

The vector $\{\sigma_1\}$ is defined thus:

$$\{\sigma_1\}^T = \sigma_{1x}, \sigma_{1y}, \sigma_{1z}, \sigma_{1xy}, \sigma_{1xz}, \sigma_{1yz} \tag{4.115}$$

Designating

$$\{\sigma_2\} = \{\sigma_0\} - \{\sigma_1\} \tag{4.116}$$

the restraint, which ensures the constant deformations in the first step of the analysis, must act on the structure by the reactions $-\{q_2\}$ which have a character of volume forces in this case. The equation of equilibrium of the member yields

$$q_{2x} = \frac{\partial \sigma_{2x}}{\partial x} + \frac{\partial \sigma_{2yx}}{\partial y} + \frac{\partial \sigma_{2zx}}{\partial z}$$

$$q_{2y} = \frac{\partial \sigma_{2xy}}{\partial x} + \frac{\partial \sigma_{2y}}{\partial y} + \frac{\partial \sigma_{2zy}}{\partial z} \tag{4.117}$$

$$q_{2z} = \frac{\partial \sigma_{2xz}}{\partial x} + \frac{\partial \sigma_{2yz}}{\partial y} + \frac{\partial \sigma_{2z}}{\partial z}$$

Proceeding by way of the approximate relaxation method, we now subject the structure, considered ideally elastic, to volume forces $\{q_2\}$; in this way, we obtain the stresses $\{\bar{\sigma}_2\}$. The stress of the structure at time t is then characterized by the stresses

$$\{\sigma\} = \{\sigma_1\} + \{\bar{\sigma}_2\} \tag{4.118}$$

The analysis of thin-walled structures is also based on the described general principle of the method. Let us consider a thin-walled structure (e.g. a shell) pictured in Fig. 4.10. The stress of the structure is characterized, for example, by

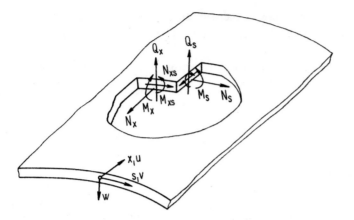

Fig. 4.10 Internal forces in a shell; the co-ordinate system used

the internal membrane forces N_x, N_s and N_{xs}, and by the slab moments and shear forces M_x, M_s, M_{xs}, Q_x and Q_s; the deformation state is described by the displacements u, v and w (Fig. 4.10). The values of these quantities at time t_0 are known.

Designating

$$\{f\}^T = N_x, N_s, N_{xs}, M_x, M_s, M_{xs}, Q_x, Q_s \tag{4.119}$$

we can write a relationship, analogous to Eq. (4.113) for the internal forces $\{f_1\}$ necessary to maintain the deformation of the structure equal to that at the beginning of the investigated process at time t_0. Also, we can determine the internal forces $\{f_2\}$ in accordance with Eq. (4.116) and we can determine the loading $\{q_2\}$ for the second step of the analysis from the equations of equilibrium of a member of the analysed structure.

If the analysed thin-walled structure is composed of plane components of generally different ages, but of uniform age for any one component (for example, slabs, walls, folded plates, thin-walled beams, etc.), the membrane and slab actions are mutually independent within their extent. The magnitude of the loading $\{q_2\}$ may be then expressed by the relations

$$q_{2x} = \frac{\partial N_{2x}}{\partial x} + \frac{\partial N_{2xs}}{\partial s}$$

$$q_{2s} = \frac{\partial N_{2xs}}{\partial x} + \frac{\partial N_{2s}}{\partial s} \tag{4.120}$$

$$q_{2z} = -\left(\frac{\partial^2 M_{2x}}{\partial x^2} + 2 \frac{\partial^2 M_{2xs}}{\partial x\, \partial s} + \frac{\partial^2 M_{2s}}{\partial s^2} \right)$$

Besides these distributed loads for the second step of the analysis, joint reactions of a force or moment character develop at the boundaries of the plane components, at the ridges and at the contacts of components having different ages. The

206

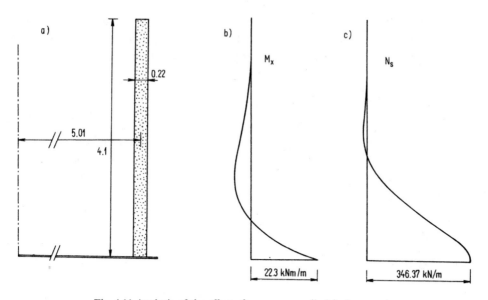

Fig. 4.11 Analysis of the effect of creep on a cylindrical reservoir
(a) cross-section of the structure; (b) diagram of longitudinal bending moments; (c) diagram of hoop normal forces

magnitude of the reactions is to be determined from the equilibrium of the ridges.

The ideally elastic structure is subjected to the loading $\{q_2\}$ so that the internal forces $\{\bar{f}_2\}$ are obtained. The required resulting stress state of the structure at time t is characterized by the internal forces $\{f\}$ which are found in accordance with Eq. (compare Eq. (4.118))

$$\{f\} = \{f_1\} + \{\bar{f}_2\} \qquad (4.121)$$

The previous conclusions, concerning the convergence of the method used for the analysis of frame structures, also apply for this case of thin-walled and three-dimensional structures, as well as the principles governing the analysis of the effect of forced deformations (for example, support settlements) and the analysis of the deformation of the structure.

It is evident that, on the basis of a general principle, one may analyse the effect of creep and shrinkage together with the effect of temperature on any frame, and on thin-walled and three-dimensional structures whose elastic analysis is known.

The versatility of the described method may be demonstrated by the example of an analysis concerning the effect of shrinkage on the stress of a cylindrical reservoir. The wall, of height 4.1 m, thickness $b = 0.22$ m and of mean radius 5.01 m (Fig. 4.11a), is monolithically connected with a rigid foundation, in which the process of shrinkage has been practically terminated at the time of casting the wall.

207

The modulus of elasticity of concrete, $E = 16\,000$ MPa, and the Poisson's ratio, $v = 0.15$. The analysis considers the safe assumption that the magnitude of shrinkage will not be reduced by the water content of the reservoir.

Hence, it is considered

$$\varepsilon_s(\infty) = 4 \times 10^{-4}$$

In the first step of the analysis, during which the deformation of the reservoir is maintained, of the internal forces $\{f_1\}$ only the specific circumferential normal force N_{1s} is non-zero due to the axial symmetry of the reservoir when the effect of shrinkage of concrete is assessed; this force is of equal magnitude for all the points of the structure and equals $0.346\,4$ MN m^{-1}. The internal forces $\{f_2\}$ again will have a non-zero component N_{2s}, which, in accordance with Eq. (4.116), is

$$N_{2s} = N_{0s} - N_{1s} = -N_{1s} = 0.346\,4 \text{ MN m}^{-1}$$

considering that, at the beginning of the investigated process, the structure was without stress (i.e. $N_{0s} = 0$). The loading $\{q_2\}$, whose non-zero component will only be the radial one q_{2z}, can be determined from the equations of equilibrium of an element of the cylindrical shell. The loading acts inside the reservoir and its magnitude is

$$q_2 = 0.346\,4/5.01 = 0.069\,1 \text{ MN m}^{-2}$$

Subjecting the elastic structure to this radial loading, we obtain the internal forces $\{\bar{f}_2\}$ and, in accordance with Eq. (4.121), we find the resulting stress state of the structure after the termination of shrinkage. The reservoir is axially symmetrical, and hence the vector of the internal forces $\{f\}$ (see Eq. (4.119)) will have the only non-zero components N_s, M_x, M_s and Q_x; the distribution of the two most significant forces along the wall of the reservoir is pictured in Fig. 4.11b, c.

The effects of creep are significant especially on three-dimensional thin-walled bridge structures whose structural system changes during the construction. To estimate the importance of this phenomenon, a simple example of the analysis of a cellular bridge structure is presented.

A continuous box-girder of three spans (30 m, 50 m and 30 m — see Fig. 4.12b) has a cross-section depicted in Fig. 4.12a. The transverse diaphragms, rigid in their own plane, are placed only at the support cross-sections, and they merely stiffen the boxes, while the cantilevers are not attached to them. The structure is cast on the centering in two stages: the left half (the dotted part of cross-section shown in Fig. 4.12a) is cast first and prestressed, after which the centering is shifted to the other half. Then, after a winter pause, the second half is cast, prestressed and connected with the first, older girder. At this instant, the age of the older girder of the structure is 9 months and that of the younger part 1 month. The purpose is to determine the stress state of the structure induced by its dead weight and due to the prestressing after 17 months after the connection, i.e. at the time

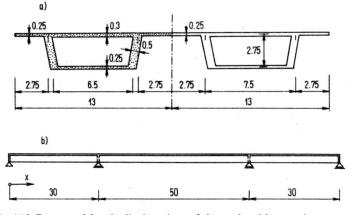

Fig. 4.12 Cross- and longitudinal sections of the analysed box-section structure

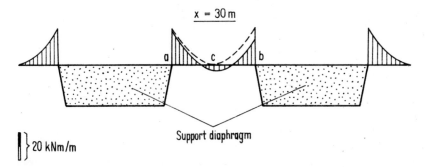

Fig. 4.13 Effect of creep on the distribution of transverse bending moments at the support cross-section of the analysed structure

when the age of the older concrete is 26 months and that of the younger concrete is 18 months.

At the instant of connection of both the cantilevers, the two halves of the structure are acting independently and their stresses induced by dead weight and prestressing are equal. No transverse bending moment or shear force develop in the connection and the stress of the slab connecting the cells remains the same as that of the opposite cantilevers. The distribution of the transverse bending moments at the cross-sections over the intermediate supports ($x = 30$ m) and at the mid-length of the structure ($x = 55$ m) is indicated by the dashed lines of Fig. 4.13 and Fig. 4.14, respectively, for the instant of connection.

The stress of the structure will change in time due to creep which causes a stress redistribution from the younger to the older girder. Thus, for example, Table 4.2 shows the distribution of the longitudinal bending moments induced by the dead weight of the girder and of the effects of prestressing on both components of the structure at the end of the investigated time interval (17 months after the connection).

209

Table 4.2 Redistribution of bending moments taken by the individual box girders

x [m]	Note	Total bending moment [MN m]	Bending moment taken by		Ratio
			older girder	younger girder	
12		4.9552	2.4528	2.5024	49.5 : 50.5
30	intermediate support	−22.7798	−12.0293	−10.7505	52.8 : 47.2
55	mid-point of 2nd span	18.4638	9.9016	8.5622	53.6 : 46.4

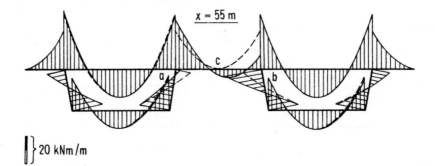

Fig. 4.14 Effect of creep on the distribution of transverse bending moments at the cross-section at mid-length of the analysed structure

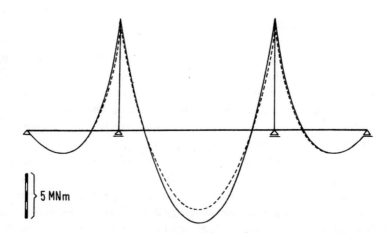

Fig. 4.15 Distribution of bending moments taken by the individual box-section girders, 17 months after the structure has been made monolithic

The distribution of the bending moments carried at this time by the older girder is shown in Fig. 4.15 (solid line); a dashed line depicts the bending moment diagram in the younger girder.

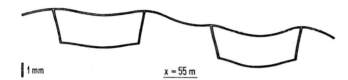

1 mm   x = 55 m

Fig. 4.16 Different deflections and distortions of the shape of the cross-section of a structure

Table 4.3   Normal forces acting in the individual box girders due to shear flow in the connecting slab

x [m]	Normal force [kN]	
	older girder	younger girder
12	− 42.8	42.8
30	88.0	− 88.0
55	− 145.4	145.4

Due to the non-uniformly increasing deflection (this increase of deflection and a distortion of the cross-section at mid-point of the 2nd span due to the connection of the structure are shown in Fig. 4.16), where the younger girder tends to deflect more, the cooperation of both components is ensured not only by the flexural rigidity of the connecting slab, but also by its in-plane-membrane-rigidity. In this way, the stress of a membrane character develops in the slab, and the shear stresses resulting from it are transferred into the box girders as a longitudinal continuously distributed shear flow, whose consequence is the stressing of the box girders by normal forces. The values of these normal forces at some cross-sections are listed in Table 4.3. If the additional normal force stressing the cross-section of one component is a tensile force, the other component is stressed by a compressive force of equal magnitude, because the total additional normal force acting at the cross-section as a whole must equal zero.

The transverse stressing of the structure, characterized by the transverse bending moments, drawn by a solid line in Figs. 4.13 and 4.14 for the cross-sections over the intermediate support and at the mid-point of the structure, respectively, is influenced by creep especially in the zone of the slab connecting both box girders. If, at the instant of connection, this slab was stressed like two opposite cantilevers with a bending moment at the clamping equal to 24.58 kN m/m and with zero value of the bending moment at mid-span of the connecting slab, then creep affects these

211

transverse bending moments so that, at the end of the investigated time interval (17 months after the connection), they attain the values shown in Table 4.4 (compare with Figs. 4.13 and 4.14) at the clamping of the connecting slab and at mid-span of this slab, respectively.

Table 4.4 Redistribution of transverse bending moments in the connecting slab due to creep

x [m]	[kN m/m]			$^aM_s : {}^bM_s$
	^aM_s	^cM_s	^bM_s	
12	25.06	4.10	15.90	61.18 : 38.82
30	23.60	4.38	16.80	58.41 : 41.59
55	31.81	4.31	8.73	78.50 : 21.50

The significance of the redistribution of the stresses from the younger into the older girder is evident from Table 4.4 and from Figs. 4.13 and 4.14; apart from that, a zone of positive transverse bending moments develops near the mid-span of the connecting slab. The connecting point, which was without transverse stress at the instant of connection, now has to carry a rather large transverse bending moment ^cM_s and a shear force resulting from the different values of the moments ^aM_s and ^bM_s.

The results of the analysed example prove the importance of studying the effects of creep also on three-dimensional thin-walled bridge structures. For the sake of illustration of the described method, the behaviour of the structure was analysed merely in the time interval of 17 months after the connection of the structure; if the entire service life of the bridge were considered (for $t \to \infty$), the effect of creep would be still more pronounced and its significance for the analysis of structures cast from concrete of different age, or for a change in the structural system, would be irrefutable.

4.3 Analysis of slender structural members

The behaviour of compressed slender concrete members can be substantially (and unfavourably) affected by creep. Even though the load acting on the deflected (or curved) member remains constant, the deformations produced by creep increase the load effects and thus induce additional deformations of the member. The analysis of slender members can therefore not be accomplished by merely increasing the deflection by the ratio of effective and elasticity moduli.

Generally, the strain of a concrete member at time t is

$$\varepsilon(t) = \frac{\sigma(t_0)}{E(t_0)}[1 + \varphi(t, t_0)] + \int_{t_0}^{t} \frac{d\sigma(\tau)}{d\tau}\left[\frac{1}{E(\tau)} + \frac{\varphi(t, \tau)}{E(\tau)}\right]d\tau \qquad (4.122)$$

where $\varphi(t, \tau)$ is the creep coefficient of concrete (also see Eq. (2.22)).

By the known conversion (integration by parts) (see Eqs. (2.23) and (2.25)), this equation can be transformed into the simple form

$$\dot{\varepsilon}(t) = \frac{\sigma(t)}{E(t)} - \int_{t_0}^{t} \sigma(\tau)\frac{\partial}{\partial\tau}\left[\frac{1}{E(\tau)} + \frac{\varphi(t, \tau)}{E(\tau)}\right]d\tau \qquad (4.123)$$

(see Eq. (2.38)).

On the basis of the usual assumptions about the stress distribution in cross-sections of members subjected to bending, the curvature of the member deflection is

$$\frac{\partial^2 y(x, t)}{\partial x^2} = -\frac{M(x, t)}{E(t)I} +$$

$$+ \int_{t_0}^{t} M(x, \tau)\frac{\partial}{\partial t}\left\{\frac{1}{E(\tau)I}[1 + \varphi(t, \tau)]\right\}d\tau \qquad (4.124)$$

where the moment M is determined by the magnitude of the load and its position on the deformed system. For example, for a member of length l, with hinges at the ends and compressed by a time-variable force $P(t)$, with initial deflections in the form $y_0 = i \sin\frac{\pi x}{l}$, the moment is

$$M(x, t) = P(t)[y(x, t) + y_0(x)] = P(t)[a(t) + i]\sin\frac{\pi x}{l} \qquad (4.125)$$

If Eq. (4.125) is substituted into Eq. (4.124), the required function of deflection y (or its amplitude a) appears in the integral of Eq. (4.124) and the function y is not only a function of time but also, e.g., of sustained load period; this complex problem is usually solved numerically.

So, for example, we may use the time-discretization method, whose principle is discussed in Sect. 2.3.3 and in [2.5].

Equation (4.124), after substitution of Eq. (4.125), yields

$$a(t)\frac{\pi^2}{l^2}I = \frac{P(t)[a(t) + i]}{E(t)} -$$

$$- \int_{t_0}^{t} P(\tau)[a(\tau) + i]\frac{\partial}{\partial\tau}\left\{\frac{1}{E(\tau)}[1 + \varphi(t, \tau)]\right\}d\tau \qquad (4.126)$$

and, in accordance with the principle of the time-discretization method in which the analysed time interval $\langle t_0, t\rangle$ is divided into k sub-intervals Δt $\left(\Delta t = \frac{t - t_0}{k}\right)$,

213

Eq. (4.126) may be written

$$a(t)\frac{\pi^2}{l^2}I = \frac{P(t)\left[a(t)+i\right]}{E(t)} -$$

$$- \sum_{j=0}^{k-1} \int_{t_0+j\Delta t}^{t_0+(j+1)\Delta t} P(\tau)\left[a(\tau)+i\right]\frac{\partial}{\partial\tau}\left\{\frac{1}{E(\tau)}\left[1+\varphi(t,\tau)\right]\right\}d\tau \tag{4.127}$$

Let us introduce the designation

$$a(\tau) = a(t_0 + j\,\Delta t) = a_j$$

$$P(\tau) = P(t_0 + j\,\Delta t) = P_j \tag{4.128}$$

Equation (4.127) then reads

$$a(t)\frac{\pi^2}{l^2}I = \frac{P(t)\left[a(t)+i\right]}{E(t)} + \sum_{j=0}^{k-1} P_j(a_j+i)\,F_j \tag{4.129}$$

where the coefficient F_j is given by Eq. (2.42) or Eq. (2.43).

If we designate that

$$\frac{\pi^2}{l^2}E_jI = P_{Ej} \tag{4.130}$$

which has the meaning of Euler's load for the j-th time interval, and if Eq. (4.129) is written for the successively growing k, we obtain

$$\begin{bmatrix} P_{E0}-P, & 0, & 0, & \\ -P_0E_1F_0, & P_{E1}-P_1, & 0, & \\ -P_0E_2F_0, & -P_1E_2F_1, & P_{E2}-P_2, & \\ \vdots & \vdots & \vdots & \\ -P_0E_kF_0, & -P_1E_kF_1, & -P_2E_kF_2, & P_{Ek}-P_k \end{bmatrix} \begin{Bmatrix} a_0 \\ a_1 \\ a_2 \\ \vdots \\ a_k \end{Bmatrix} = i \begin{bmatrix} 1, & 0, & 0, & \\ E_1F_0, & 1, & 0. & \\ E_2F_0, & E_2F_1, & 1, & \\ \vdots & \vdots & \vdots & \\ E_kF_0, & E_kF_1, & E_kF_2, & \dots 1 \end{bmatrix} \begin{Bmatrix} P_0 \\ P_1 \\ P_2 \\ \vdots \\ P_k \end{Bmatrix}$$

or, in a concise form:

$$[U]\{a\} = i\,[V]\,\{P\} \tag{4.131}$$

Equation (4.131) is a system of algebraic equations for the amplitudes of the deflection in the j-th time sub-interval. Considering that the matrix $[U]$ of this system is already triangularized, the individual unknowns (starting with the first, a_0) may be calculated successively from the individual equations. This is a very simple operation, manageable with the aid of pocket calculators.

The first Eq. (4.131) evidently yields

$$a_0 = \frac{P_0 i}{P_{E0}-P_0} \tag{4.132}$$

which is the amplitude of the deflection at time t_0, representing the initial condition of the time development of the deformation of the member. It is a familiar relation for the elastic deflection of a slender member with imperfections, usually

214

written in the form

$$a_0 = \frac{i}{\dfrac{P_{E0}}{P_0} - 1} \tag{4.\overline{132}}$$

If we consider a member without imperfections ($i = 0$), i.e. a member ideally straight, Eq. (4.131) allows the study of the stability of the member. In other words, the values of the forces P_j for which the member maintains its deformation can be found, i.e. at which $a_j \neq 0$. If it is required, for example, that $a_0 = a_1 = a_2 = \ldots = a$, then we obtain successively

$$\left.\begin{aligned} P_{0,\mathrm{cr}} &= P_{E0} &\text{*)} \\ P_{1,\mathrm{cr}} &= P_{E1} - P_{E0}E_1F_0 \\ \vdots \quad &\quad \vdots \qquad \vdots \end{aligned}\right\} \tag{4.133}$$

To estimate the studied phenomenon, let us consider the force P to be constant in time with the approach based on the rate-of-creep theory. However, it must be noted again that the use of the rate-of-creep theory is principally incorrect and leads to an underestimation of this phenomenon as it does not take into account delayed elastic deformations, which are not dependent on the age of concrete, but on the effects of the sustained load period (i.e. even old concrete newly loaded is subject to long-term deformation). Consequently, the following results can only be utilized for an assessment of the deformations of a member made of young concrete in a short time interval (see also [4.2]).

The only advantage of this procedure is that Eq. (4.124) can be transformed to the differential equation of the form

$$\frac{\partial^3 y(x, t)}{\partial^2 x\, \partial t} = -\frac{1}{EI}\left[\frac{\partial M(x, t)}{\partial t} + M(x, t)\frac{d\varphi(t)}{dt}\right] \tag{4.134}$$

By introducing Eq. (4.125), a common differential equation for the amplitude a ($P = $ constant) is obtained:

$$\frac{da(t)}{dt}\left(\frac{P_E}{P} - 1\right) - a(t)\frac{d\varphi(t)}{dt} = i\frac{d\varphi(t)}{dt} \tag{4.135}$$

whose solution at the initial condition (expressing elastic deflection)

$$a(t_0) = \frac{i}{\dfrac{P_E}{P} - 1} \tag{4.136}$$

---

*) The state at the beginning of the phenomenon at time t_0, in accordance with Eq. (4.130), is expressed by the well known Euler's relation

$$P_{0,\mathrm{cr}} = P_{E0} = \frac{\pi^2}{l^2}E_0 I$$

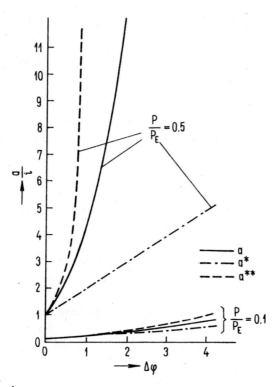

Fig. 4.17 Deflection values given by different formulae

is

$$a(t) = i \left(\frac{1}{1 - \dfrac{P}{P_E}} \, e^{\frac{\Delta\varphi}{\frac{P_E}{P} - 1}} - 1 \right) \qquad (4.137)$$

where $\Delta\varphi$ is the difference of the creep coefficient values in the analysed time interval (compare with Eq. (4.103)).

By merely multiplying the elastic deflections by the ratio of the effective and elasticity moduli, the following relation would be obtained

$$a^*(t) = \frac{i}{\dfrac{P_E}{P} - 1} (1 + \Delta\varphi) \qquad (4.138)$$

which does not reflect the long-term effects of the increasing deflection with time; the result is an underestimation of the effect of the phenomenon.

On the other hand, by assuming that the final force effects were already acting since the beginning of this phenomenon, the following formula would be obtained:

$$a^{**}(t) = \frac{i}{\dfrac{P_E}{P(1 + \Delta\varphi)} - 1} \qquad (4.139)$$

which overestimates the studied phenomenon.

216

In Fig. 4.17, the curves of the ratio $\frac{a}{i}$ (i.e. ratio of deflection and initial imperfection) are plotted against the difference of the creep coefficients, expressing, according to the rate-of-creep theory, the time interval of the sustained load effects. The results, according to Eq. (4.137), fall between the values given by Eqs. (4.138) and (4.139); the differences for an increasing load are very significant and indicate the errors in Eq. (4.138).

In can be seen that even the inaccurate rate-of-creep theory indicates the considerable significance of creep. Its application, however, is only useful for an assessment of the increase of deflections in structures made from young concrete, and not for the determination of the critical loading of a member with imperfections, since it is clear that, according to this theory, Eq. (4.137) would again result in $P_{cr} = P_E$. By starting from the illustrative Eq. (4.122) and/or Eq. (4.123), an accurate picture of the behaviour of the member would be obtained, and the allowable load which is dependent on the time of its introduction and on its duration could be defined; also, the period during which the member is capable of transferring the given load, could be possibly determined.

217

5

Structural members and structures in which cracks may develop

Cracks develop in the tensioned parts of the cross-sections of reinforced concrete beams subjected to service load; this affects the cross-sectional rigidity which varies from one section to another. An example of the behaviour of a reinforced concrete member, subjected to continuous load, is shown in Fig. 5.1.

As long as the bending moment M_v of the external forces does not reach the value of the moment M_F at which cracks develop, the entire cross-section is effective and it behaves more or less plastically mainly in the tensioned zones which are designated by number I in Fig. 5.1. The displacement of the neutral axis x, the alternation of the curvature ω and of the cross-sectional rigidity B_i correspond to this behaviour; the neutral axis moves towards the compressed edge of the cross-section, the curvature of the cross-section increases with the increasing

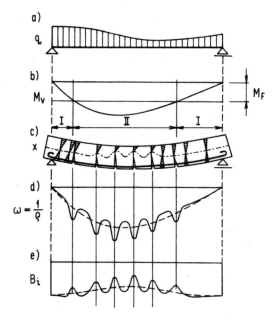

Fig. 5.1 Neutral axis, curvature and stiffness of a member with cracks

218

bending moment and with the progressive plasticization of the cross-section, and the rigidity of the cross-section decreases, so that

$$B_i = \frac{M}{\omega}$$

Cracks develop in the zones II, but between the cracks the cross-sections act fully. In sections with cracks, the neutral axis moves significantly towards the compressed edge (Fig. 5.1c), so that, at sections with larger bending moments, the stress diagram of compressed concrete may be curvilinear even under service conditions. Owing to the cracks, the tension below the neutral axis is taken mostly by steel, but also a small part of concrete is stressed by tension just below the neutral axis. In the sections between the cracks, the neutral axis moves downwards; its position, however, depends on the extent of displacement of steel reinforcement in concrete.

Obviously, the state of stress of a cross-section and its strain are influenced by many factors of different significance. They are: the static arrangement of the member, the type of loading (continuous load, point loads), the shape of the cross-section, the distribution of the reinforcement along the member, the steel ratio (or the degree of prestressing), the deformation properties of concrete and steel, the adhesion between the reinforcement and concrete, and last, but not least, the creep and shrinkage of concrete in the compressive and tensile parts of the cross-section.

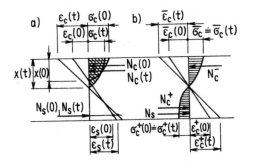

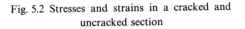

Fig. 5.2 Stresses and strains in a cracked and uncracked section

Let us assume that cracks, caused by the service load, develop in a structural member subject to a bending moment which is constant in time. Then, assuming that the strain of the cross-section in the crack is proportional to its distance from the neutral axis, the contraction of concrete at the upper edge of the cross-section in time $t \to 0$ is $\varepsilon_c(0)$ and the elongation of the reinforcement equals $\varepsilon_s(0)$ (Fig. 5.2a); the distance of the neutral axis is $x(0)$ from the upper edge of the cross-section and the stress of concrete in the outermost fibres is $\sigma_c(0)$ at the same instant. The effect of creep is to increase the strain of the outermost fibre from $\varepsilon_c(0)$ to $\varepsilon_c(t)$ and, simultaneously, $\varepsilon_s(0)$ increases to $\varepsilon_s(t)$. In consequence, the neutral axis moves downwards and its distance from the upper edge will equal $x(t)$. This will cause a new stress distribution in the compressed part of the concrete; the stress will decrease from $\sigma_c(0)$ to $\sigma_c(t)$, because the moment caused by the internal forces must be equal to that caused by the external forces which does not change in time. The lever arm of the internal forces decreases to some extent and the force acting in the reinforcement increases from $N_s(0)$ to $N_s(t)$. Also, the curvature at this section increases.

At the sections without cracks (Fig. 5.2b), the strain of both the compressed and tensioned parts of the cross-section is increased due to creep. Assuming a linear relationship between creep and stress, and identical creep behaviour in the compressed and tensioned concrete, while neglecting the small influence of reinforcement, we may consider that the stress at a cross-section subjected to a long-term loading will be constant so that its distribution and magnitude will be the same at any time (Fig. 5.2b); the curvature of the cross-section at time t will increase relative to that at time $t = 0$.

The tests of reinforced structural members with cracks [5.1, 5.2] subjected to a long-term load at the level of the service load indicate that the increase of stress in the reinforcement is not too large, and thus it is estimated that, due to creep and the effect of cracks, it never increases more than by 15 per cent. According to [5.3], for example, this increase equals 5 to 15 per cent in members subjected to a bending moment. On the other hand, the decrease of stress $\sigma_c(t)$ in the compressed part of the concrete may be rather large.

The development of cracks caused by a long-term load in beams subjected to bending has not been explained until now. The reported tests [5.3] lead to the conclusion that cracks increase in width due to creep and, moreover, new cracks develop even if the load is maintained within the limits of the service load. This finding disagrees with the conclusion about the increase of the area of the compressed part of concrete under a long-term loading; however, it should be noted that, in this problem, the effects of shrinkage and of the displacement of the reinforcement in the concrete are significant, the latter effect also having a long-term character. Evidently, the explanation of the causes of this phenomenon is very complex and the proposed methods of the analysis of creep in reinforced-concrete members should be approached with this aspect in mind.

Shrinkage alone may lead to the formation of cracks in reinforced members without load, because the reinforcement restrains the free development of shrinkage strain; then, compressive stress develops in the reinforcement, but concrete, on the other hand, is subjected to tensile stress. Large effects are to be expected in asymmetrically reinforced members, especially in a simply reinforced cross-section where tension in concrete develops in the zones close to the reinforcement and compression appears in the parts without reinforcement; thus, the member is subjected to bending.

If the tensile stress does not exceed the limits at which cracks develop, the cross-section can be analysed in agreement with Sect. 4. On the other hand, if a crack develops in a cross-section, the state of stress in steel and concrete changes abruptly at this section, but the original elastic state of stress is maintained in the uncracked portions of the member. Consequently, another method of analysis must be employed.

The analyses of the effects of creep and shrinkage in reinforced concrete structural members are based on experimental investigations; consequently, the various methods of analysis differ considerably and they yield different results. A theoretically very elaborate method has been developed in the USSR, but the present situation is that there are not sufficient data available to determine all parameters exactly, and, as yet, they have not been introduced into the Codes.

Subsequently, we shall refer to those necessary parameters without which a more accurate analysis is impossible; we shall also demonstrate some methods of the analysis of the effects of long-term loading and shrinkage.

All the following analyses will be concerned with the effect of long-term influences of creep and shrinkage on the redistribution of stresses and on the rigidity of the structural member.

5.1 Deformation and rigidity of structural members with cracks

5.1.1 Factors influencing strain

The analyses of deformation and the calculations of rigidity of reinforced concrete members, subjected to bending and to a compressive or tensile force, are usually based on the assumption of a triangular or rectangular stress diagram in the compressed concrete at the section with cracks, the latter shape of the stress block being more common at the present time. In the analyses, it is also necessary to consider the difference in the strains of concrete and steel at a section with and without cracks. For this reason, both the stress-block problem and the influence of the difference in strain of concrete and steel at short- and long-term loadings will be discussed.

It should be noted that the subsequent discussed problems are still being elaborated and further results of experimental investigation are needed to make the values of the above mentioned factors more accurate. Hence, the conclusions of the following Sections should be regarded with caution.

5.1.1.1 *Coefficient of compressive stress distribution $\bar{\omega}$ and strain coefficient $\bar{v}$ at a section with cracks*

In the service state, the stress diagram in the compressed concrete of a cross-section subjected to bending is approximately triangular, especially in the case of short-term loading. If x designates the position of the neutral axis, b the width of the cross-section and if the diagram of the compressive stress is triangular with the stress $\bar{\sigma}_c$ in the outermost fibre, then the force in the concrete under compression is

$$N_c = 0.5xb\bar{\sigma}_c$$

If the stress $\bar{\sigma}_c$ increases, the linear stress distribution changes into a curvilinear one, and the ratio of the elastic strain ε_e of concrete under compression and of the total strain ε_t (i.e., of the elastic strain ε_e and the plastic strain ε_p) also changes; this change strongly depends on the duration of the loading. We shall designate this ratio

$$\bar{v} = \frac{\varepsilon_e}{\varepsilon_e + \varepsilon_p} = \frac{\varepsilon_e}{\varepsilon_t} \tag{5.1}$$

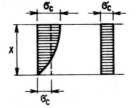

Fig. 5.3 Curvilinear and constant shape of the stress block

If the distribution of the compressive stress σ_c is considered rectangular in the analysis, the change in stress distribution has to be expressed with the aid of a coefficient $\bar{\omega}$; obviously, the resulting forces in the concrete under compression must be equal in either case, and hence, we have (see Fig. 5.3)

$$\bar{\omega}xb\bar{\sigma}_c = xb\sigma_c$$

and

$$\bar{\omega} = \frac{\sigma_c}{\bar{\sigma}_c} \tag{5.2}$$

The coefficient $\bar{\omega}$ equals 0.5 for a triangular stress distribution and it increases with the increasing load, while, for rectangular stress distribution, $\bar{\omega} = 1$.

If the modulus of elasticity $E_c = \dfrac{\bar{\sigma}_c}{\varepsilon_e} = \dfrac{\sigma_c}{\bar{\omega}\varepsilon_e}$, the modulus of deformability

$E_{c0} = \dfrac{\sigma_c}{\varepsilon_t}$, and $\varepsilon_e = \bar{v}\varepsilon_t$ (in accordance with Eq. 5.1), then the modulus of deformability is

$$E_{c0} = \bar{\omega}\bar{v}E_c = vE_c \qquad (5.3)$$

where the product of the coefficients $\bar{\omega}$ and $\bar{v}$ is designated by the symbol v.

For a short-term service load, the value of the coefficient $\bar{v}$ approaches unity, and since the stress at this loading state has an approximately triangular distribution (i.e. $\bar{\omega} = 0.5$), then $v = 0.50$. If the stress of concrete σ_c increases, the decrease of $\bar{v}$ is more rapid than the increase of $\bar{\omega}$, and hence, the value of v diminishes. Usually, the coefficient v is considered more or less constant [5.5] for short-term loading and for the cases when the ratio of the stress to the compressive strength in bending (σ_c/R_{cb}) is less than 0.6 to 0.7. However, some tests [5.7] have indicated that this coefficient varies even below this limit; this is shown in Fig. 5.4 which depicts the distribution of the coefficients $\bar{v}$ and $\bar{\omega}$. The Standards give the following values of v for short-term loading:

for reinforced concrete members subject to simple bending, $v = 0.5$,

for other cases, $v = 0.45$.

In some Codes, $E_{c0} = 0.5E_c$ is stated, i.e. $v = 0.5$ for all cases. According to test results [5.4], the value of the coefficient $v(t)$ decreases in time under long-

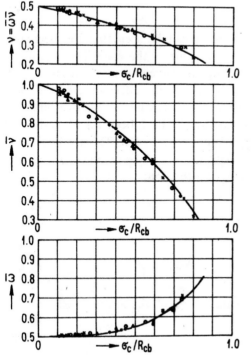

Fig. 5.4 Coefficients v, $\bar{v}$ and $\bar{\omega}$

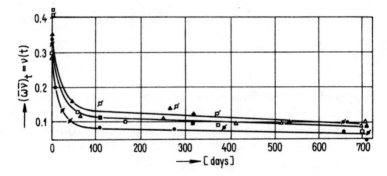

Fig. 5.5 Variation of the coefficient *v* with time (dots indicate the test results)

Table 5.1   Coefficient $v(t)$ for long-term loading

Ambient conditions	very humid	humid	dry	very dry
Average annual relative humidity [%]	80	60 to 80	40 to 60	40
SN $v(t)$	0.20	0.15	0.15	0.10
ČSN 73 1201 $v(t)$	0.19	0.15	0.12	0.11

term loading as shown in Fig. 5.5 [5.4]; here it can be seen that the decrease is very large: after approximately 170 days, the value of the coefficient is within $\langle 0.175, 0.26 \rangle$ and, after two years of loading, it decreases to values between 0.1 and 0.05 in some cases. This decrease is due to the effect of creep in the compressed part of the cross-section, and hence, it also depends on the hygrometric ambient conditions. Accordingly, the following values have been introduced into the Czechoslovak (ČSN) and Soviet (SN) Standards [5.5] (see Table 5.1).

The coefficient *v* may also be calculated with the aid of the creep coefficient φ; the modulus of deformability is expressed by the relation:

$$E_c(t) = 0.5 \frac{1}{1 + \varphi} E_c \qquad (5.3a)$$

which yields

$$v(t) = 0.5 \frac{1}{1 + \varphi} \qquad (5.3b)$$

In this case, the formula used for the calculation of the creep coefficient is

$$\varphi = \beta \frac{\delta}{\vartheta} \qquad (5.4)$$

224

Table 5.2   Values of the hygrometric coefficient δ

Ambient conditions	watery	very humid	humid	dry	very dry
Relative humidity [%]	100	> 80	60 to 80	40 to 60	< 40
δ	1.00	1.50	2.30	2.85	3.30

This formula is provided in the Standard: ČSN 73 1201, where
β = coefficient expressing the age of concrete and the duration of loading, as
determined from the expression (Dischinger's form, see Sect. 2)

$$\beta = 1.2[(1 - e^{-1.6t}) - (1 - e^{-1.6t_0})] \qquad (5.5)$$

δ = coefficient of hygrometric conditions, as given in Table 5.2;
ϑ = ratio of the strength of concrete at the time of the application of load to the
given strength of concrete.

In Eq. (5.5), t_0 and t designate the ages of concrete in years at the commencement
and at the end of the time interval, respectively, in which the effect of creep is
investigated.

Hence, if a structure is loaded when $t_0 = 28$ days, then $\vartheta = 1$ and, for $t \to \infty$,
the coefficients v will have the values given in the bottom line of Table 5.1,
depending on storage condition.

It should be noted that the actual values of the coefficient v may differ largely
from those calculated and that more accurate expressions for the calculation of
this coefficient do not represent an improvement in the agreement with reality.
This applies to the coefficient v under short-term loading and, more so, to its
values under long-term loading.

5.1.1.2 Coefficient ψ_s

The coefficient ψ_s expresses the ratio of the average strain (or stress) of the
tensile reinforcement in a section without cracks to the strain (or stress) of the
reinforcement in a section with cracks (Fig. 5.6b$_1$); hence, it can be written

$$\psi_s = \frac{\varepsilon_s'}{\varepsilon_s} = \frac{\sigma_s'}{\sigma_s}$$

This coefficient depends on time, as it varies with the duration of loading.
Figure 5.6 shows that the coefficient ψ_s may also be calculated by means of the
strain areas of steel, so that

$$\psi_s = \frac{\varepsilon_s l_t - \omega_0 \varepsilon_{s1} l_t}{\varepsilon_s l_t} = 1 - \omega_0 \frac{\varepsilon_{s1}}{\varepsilon_s}$$

225

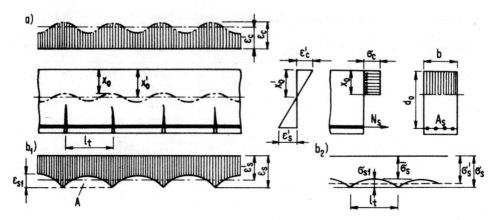

Fig. 5.6 Strains and stresses in a concrete member with cracks

and also

$$\psi_s = 1 - \omega_0 \frac{\sigma_{s1}}{\sigma_s}$$

The symbol ω_0 expresses the ratio of the strain area A to the area $\varepsilon_{s1} l_t$; if, for example, the area is parabolic of the second degree, then $\omega_0 = 2/3$.

If a moment M acts at a section without cracks and if the part M_c of it is carried by concrete and part M_s by steel, then (with the same lever arm z inside and outside the crack (Fig. 5.6b$_2$))

$$M = \bar{M}_s + M_c = A_s \bar{\sigma}_s z + M_c = A_s(\sigma_s - \sigma_{s1}) z + M_c$$

If the same bending moment M acts at a section with a crack, then $M = A_s \sigma_s z$ and the stress is

$$\sigma_{s1} = \frac{M_c}{A_s z}$$

and consequently $\left(\text{at } \sigma_s = \dfrac{M}{A_s z} \right)$

$$\psi_s = 1 - \omega_0 \frac{M_c}{M} \tag{5.6a}$$

The ratio of the moment M_c carried by concrete to the moment M_{cF} which concrete can withstand at the onset of cracking is designated by $\chi = \dfrac{M_c}{M_{cF}}$; substituting this ratio into Eq. (5.6a), the coefficient ψ_s becomes

$$\psi_s = 1 - \omega_0 \chi \frac{M_{cF}}{M} \tag{5.6b}$$

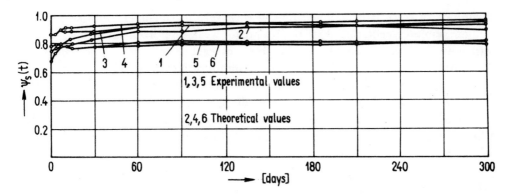

Fig. 5.7 Variation of the coefficient $\psi_s(t)$ with time: 1, 3, 5 represent the test results; 2, 4, 6 are the theoretical values

The test results [5.5] under short-term loading indicate that the magnitude of the coefficient ψ_s depends mainly on the ratio M_{cF}/M, while the product $\omega_0\chi$ may be considered constant.

The coefficient ψ_s may then be calculated by using, for example, the experimental relationship for members subjected to bending

$$\psi_s = 1.3 - \lambda \frac{M_{cF}}{M} \tag{5.7}$$

which is applicable within the limits $\langle 1.3 - \lambda \leq \psi_s \leq 1 \rangle$.

The coefficient λ depends on the surface texture of the reinforcing bars, the following values being appropriate:

$\lambda = 1.00$ for smooth steel bars;
$\lambda = 1.10$ for shaped steel bars.

Under long-term loading, the coefficient $\psi_s(t)$ also depends on time. With reference to test results [5.6], its average value increases by 10 to 20 per cent; the results of measurements of $\psi_s(t)$ from one series of these tests are represented in Fig. 5.7 in which simply and doubly reinforced concrete specimens of dimensions 10 by 20 cm were investigated. The tests yielded the relationship

$$\psi_s(t) = \psi_s(1 + \delta_1\varphi)$$

where φ is the creep coefficient for the time at which $\psi_s(t)$ is calculated, and δ_1 is an experimental quantity determined by the relation

$$\delta_1 = \frac{\psi_s(\infty) - \psi_s}{\psi_s\varphi_\infty}$$

The values: $\psi_s(\infty)$, ψ_s and φ_∞ are determined experimentally. Experiments with different reinforced members have shown that the steel ratio also influences the

227

magnitude of $\psi_s(t)$. It is assumed that, if $\psi_s \geq 0.75$ for members with strong and medium reinforcement, then $\psi_s(t) = 1.0$, and, for weak reinforcement with $\psi_s < 0.75$, $\psi_s(t)$ may be assumed equal to $1.5\psi_s$ (but not exceeding 1.00). These values are recommended only for the analyses of maximum deflection [5.6].

At present, some Codes suggest the unified value $\lambda = 0.8$ for members subjected to bending into Eq. (5.7) for the long-term effect of loading, but the effect of surface texture of the reinforcing bars is not distinguished; hence, the coefficient $\psi_s(t)$ has a higher value than in the case of short-term loading.

5.1.1.3 *Coefficient ψ_c*

The non-uniform distribution of strain (and stress) in concrete under compression in the regions between cracks is expressed by the coefficient ψ_c; it characterizes the ratio of the average compressive strain of concrete in the regions between the cracks to the compressive strain of concrete in the cracked section (Fig. 5.6a), i.e.

$$\psi_c = \frac{\varepsilon_c'}{\varepsilon_c}$$

Tests of structural members subjected to short-term loading have indicated that the value of ψ_c is within the limits $\langle 0.8; 1.0\rangle$, and hence, it is introduced into the analysis by its average value of $\psi_c = 0.9$.

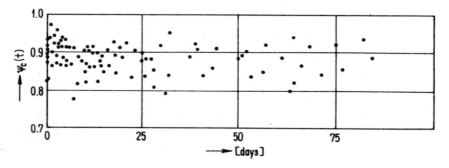

Fig. 5.8 Test results of the coefficient $\psi_c(t)$

The long-term effect of loading on the coefficient $\psi_c(t)$ has not been ascertained in detail. In Fig. 5.8, test results are highly scattered [5.7], but the coefficient does not appear to increase with time and it does not reach the value of unity in any of the experiments. Consequently, it may be assumed that the distribution of creep under compression is the same for long-term and short-term loadings.

5.1.2 The average position of the neutral axis

It has been explained that under short-term loading the neutral axis (x_0) moves closer to the compressed edge in sections with cracks than in regions without cracks (Fig. 5.6) and that, with continuation of the same load, the neutral axis recedes somewhat from this edge due to the creep of the compressed concrete; simultaneously, the reinforcing bars elongate to some extent, and the resulting neutral axis position x is determined by the effect of the strain in concrete and steel. Let us designate the new position of the neutral axis by $x(t)$. The mechanism of the displacement of the neutral axis is shown in Fig. 5.9 where the strains were observed on a member tested for a period of 274 days. The magnitude of the long-term strain clearly shows that the reinforcement strain is relatively small in relation to that of concrete. The same conclusion is derived from similar tests [5.6], as shown in Fig. 5.10; this conclusion evidently also applies after unloading of the member between approximately 280 and 290 days. Thus, the main influence which causes the displacement of the neutral axis under long-term loading is the deformation of concrete, i.e. creep.

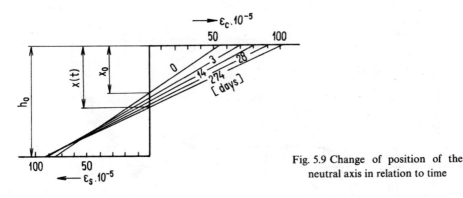

Fig. 5.9 Change of position of the neutral axis in relation to time

The determination of the positions x_0 and $x(t)$ of the neutral axis is based on the assumption of a linear relationship between the strain and its distance from the neutral axis, as shown in Fig. 5.9. *Note:* It has been found in tests that this assumption is not quite correct, both in the cases of short-term and longer-term loadings; in the latter case, the deviations from linearity are more pronounced. This is evident from the tests [5.7] (Fig. 5.11) in which strain gauges were fixed at several points to concrete under compression and tension; if only the strain of concrete under compression is considered, then non-linearity of strain of concrete near the neutral axis, as x_0 changes to $x(t)$, is evident. In the part under tension, the entire cross-section is not shown in the diagram, but only that part where a visible crack ends. Although the cross-section does not remain plane after deformation, this fact is not considered in the analysis and direct proportionality is assumed. This results in a substantial simplification of the analysis, although

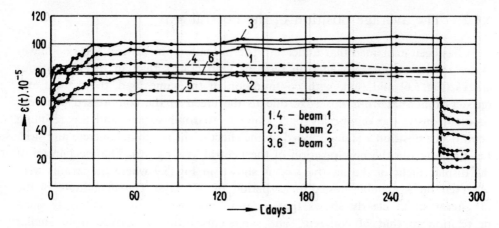

Fig. 5.10 Strain of the extreme fibre of compressed concrete (solid lines) and of tensioned steel (dashed lines) related to time;

1, 4 — beam 1; 2, 5 — beam 2; 3, 6 — beam 3

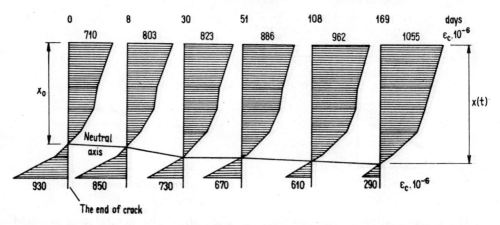

Fig. 5.11 Strain of compressed and tensioned parts of a section through crack in relation to time

it may lead to certain differences between the results and reality. Bearing in mind that other factors enter into the analysis whose values are not known sufficiently well at the moment, it seems unnecessary to complicate the calculation by attempting to express the position of the neutral axis more accurately.

The variation with time of the position of the neutral axis in a cracked section is pronounced, as shown in the graph of Fig. 5.12, which portrays the distribution of the ratio $\xi(t) = \dfrac{x(t)}{d_0}$ for 720 days on several types of reinforced concrete members subjected to bending. Here, the increase of $\xi(t)$ varies within from 35 to 40 per cent for this period of time.

230

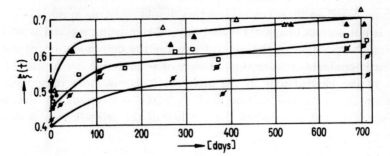

Fig. 5.12 Variation of position of the neutral axis in relation to time
(signs correspond to the results of tests)

Considering that the position of the neutral axis varies continually along the beam, one can determine its average value only; this can be accomplished from its position in a section through a crack $x(t)$ and in a section between cracks $x'(t)$. Then, the expression of the new coefficient for long-term loading would be

$$\psi_x(t) = \frac{x(t)}{x'(t)} = \frac{\xi(t)}{\xi'(t)}$$

and, for short-term loading (Fig. 5.6),

$$\psi_{x0} = \frac{x_0}{x'_0} = \frac{\xi_0}{\xi'_0}$$

The values of these coefficients are less than unity. The position of the neutral axis can be determined by means of the strains ε'_c and ε'_s by using the relation (Fig. 5.6)

$$\frac{\varepsilon'_c}{\varepsilon'_s} = \frac{x'_0}{d_0 - x'_0}$$

and, since $\varepsilon'_c = \psi_c \varepsilon_c$ and $\varepsilon'_s = \psi_s \varepsilon_s$, the coefficient ψ_{x0} for short-term loading would be

$$\psi_{x0} = \frac{x_0}{d_0}\left(1 + \frac{\psi_s \varepsilon_s}{\psi_c \varepsilon_c}\right)$$

Similarly, for long-term loading

$$\psi_x(t) = \frac{x(t)}{d_0}\left(1 + \frac{\psi_s(t)\, \varepsilon_s}{\psi_c(t)\, \varepsilon_c}\right)$$

Once the coefficients ψ_{x0} and $\psi_x(t)$ are known, the average position of the neutral axis for short-term as well as long-term loadings can be determined [5.3]; as, however, the values ψ_s and ψ_c, or $\psi_s(t)$ and $\psi_c(t)$ are approximate, the determination of the coefficients ψ_{x0} and $\psi_x(t)$ is still only approximate, so that in

231

the analyses based on these coefficients this should be borne in mind. On the other hand, it is not correct to maintain that any other existing method is accurate for calculating the position of the neutral axis for the determination of the rigidity of reinforced-concrete cross-sections. Hence, it is advisable not to use the average position of the neutral axis in the analysis; it seems more reasonable to find its position in a cracked section and to make corrections by means of the coefficients $\psi_s(t)$ and $\psi_c(t)$, when the rigidity of the cross-section is determined.

5.1.2.1 *Position of the neutral axis*

5.1.2.1.1 *Structural members subjected to bending*

First, let us demonstrate two methods which may give an idea of the procedure to be adopted in determining the position of the neutral axis in a section through a crack under long-term loading, i.e. with consideration of the time-distribution of creep; the analysis will be demonstrated on a simply reinforced rectangular cross-section with the assumption of the rate-of-creep theory. Both these methods are more time-consuming than the third method, which may be characterized as being approximate from the theoretical point of view, but is practically the most convenient and sufficiently close to reality.

All the three methods are based on the following assumptions:
(1) all cross-sections are plane even after deformation;
(2) the stress in concrete in the compressed zone is uniformly distributed;
(3) the stress in the outermost compressed fibre is

$$\sigma_c = E_{c0}\varepsilon_c$$

(4) concrete under tension is not effective;
(5) the modulus of deformability E_{c0} under short-term loading, from which the analysis starts, is

$$E_{c0} = vE_c = 0.5E_c$$

because $\bar{v} \doteq 1$ and $\bar{\omega} = 0.5$ from the preceding discussion.

(a) A more accurate analysis accounting for the effect of creep

Position of the neutral axis at time $t = 0$:
    The stress in steel is determined by the relation

$$\sigma_s = \frac{E_s}{E_{c0}} \sigma_c = n_0 \sigma_c \tag{5.8}$$

where E_s designates the modulus of elasticity of steel.

232

For the determination of x_0 we can use the strain condition at a section through crack

$$\varepsilon_s = \varepsilon_c \frac{d_0 - x_0}{x_0} \tag{5.9a}$$

and the force condition

$$N_c - N_s = 0$$

which can be also written

$$\sigma_c x_0 b - A_s \sigma_s = 0$$

Substituting from Eq. (5.8), considering Eq. (5.9a) and designating $x_0 = \xi_0 d_0$ (d_0 is the effective depth of the section), the expression for the position of the neutral axis acquires the form

$$\xi_0 = \frac{n_0 \mu}{2} \left(-1 + \sqrt{1 + \frac{4}{n_0 \mu}} \right) \tag{5.10}$$

Position of the neutral axis at instant t:

The conditions of equilibrium at an arbitrary instant t can be written

$$bx(t) \, \sigma_c(t) - A_s \sigma_s(t) = 0$$

$$bx(t) \, \sigma_c(t) \left(d_0 - \frac{x(t)}{2} \right) - M = 0$$

In this case, it is assumed that the bending moment M of the external forces is constant in time.

The strain condition (Eq. (5.9a)) changes to

$$\varepsilon_s(t) = \varepsilon_c(t) \frac{d_0 - x(t)}{x(t)} \tag{5.9b}$$

and also

$$\varepsilon_c(t) = \frac{\sigma_s(t)}{E_s} \frac{x(t)}{d_0 - x(t)} \tag{5.11}$$

Applying the rate-of-creep theory, the strain caused by creep for the time interval $\langle 0, t \rangle$ can be expressed by the relation

$$\varepsilon_c(t) = \frac{\sigma_c(t)}{E_{c0}} (1 + \varphi(t)) + \int_0^t \frac{d\sigma_c(\tau)}{d\tau} \left(\frac{1}{E_c(\tau)} + \frac{\varphi(t) - \varphi(\tau)}{E_{c0}} \right) d\tau \tag{5.12}$$

The modulus of elasticity under short-term loading E_{c0} should be considered as the starting point in agreement with the third initial assumption. Equation (5.12) is obtained from Eq. (2.21) in which the function (2.7) is used. The instant $\tau_1 = 0$ is considered in the arrangement of the equation,

Equation (5.12) expresses the variability of stress and the variation of the modulus of elasticity of concrete under compression with time. A comparison of Eqs. (5.11) and (5.12) yields the following:

$$\frac{\sigma_c(t)}{E_{c0}}(1 + \varphi(t)) + \int_0^t \frac{d\sigma_c(\tau)}{d\tau}\left(\frac{1}{E_c(\tau)} + \frac{\varphi(t) - \varphi(\tau)}{E_{c0}}\right)d\tau =$$

$$= \frac{\sigma_s(t)}{E_s}\frac{x(t)}{d_0 - x(t)}$$

Differentiation with respect to t yields (Eq. (2.64))

$$\frac{1}{E_s}\frac{d}{dt}\left(\sigma_s(t)\frac{x(t)}{d_0 - x(t)}\right) = \frac{1}{E_c(t)}\frac{d\sigma_c(t)}{dt} + \frac{\sigma_c(t)}{E_{c0}}\frac{d\varphi(t)}{dt} \qquad (5.13)$$

From the conditions of equilibrium

$$\sigma_c(t) = \frac{M}{bx(t)\left(d_0 - \dfrac{x(t)}{2}\right)} \qquad (5.14a)$$

$$\sigma_s(t) = \frac{M}{A_s\left(d_0 - \dfrac{x(t)}{2}\right)} \qquad (5.14b)$$

and, if the calculated stresses are introduced into Eq. (5.13) with $\xi(t)\,d_0$ substituted for $x_0(t)$, we obtain

$$\left[\frac{1}{n_0\mu}\frac{(2 - \xi^2(t))\,\xi(t)}{(2 - \xi(t))\,(1 - \xi(t))^2} + 2\frac{E_{c0}}{E_c(t)}\frac{1 - \xi(t)}{\xi(t)\,(2 - \xi(t))}\right]d\xi = d\varphi$$

This differential equation can be solved numerically; if, however, $E_c(t) = E_{c0} = $ = constant, which can be admitted in practical cases, a solution in closed form may be obtained. The equation then acquires the form

$$\left[\frac{1}{n_0\mu}\frac{(2 - \xi^2(t))\,\xi(t)}{(2 - \xi(t))\,(1 - \xi(t))^2} + 2\frac{1 - \xi(t)}{\xi(t)\,(2 - \xi(t))}\right]d\xi = d\varphi \qquad (5.15)$$

In the last two equations, $n_0 = \dfrac{E_s}{E_{c0}}$ and $\mu = \dfrac{A_s}{b\,d_0}$.

The solution of Eq. (5.15) yields

$$\left(\frac{4}{n_0\mu} + 1\right)\ln(2 - \xi(t)) + \frac{1}{n_0\mu}\frac{1}{1 - \xi(t)} + \frac{1}{\mu n_0}\xi(t) +$$

$$+ \ln \xi(t) = \varphi(t) + A \qquad (5.16)$$

The integration constant A is determined from the condition that, for $t = 0$,

$\varphi(t) = 0$ and $\xi(t) = \xi_0$; then

$$A = \left(\frac{4}{n_0\mu} + 1\right)\ln(2 - \xi_0) + \frac{1}{n_0\mu}\frac{1}{1 - \xi_0} + \frac{1}{\mu n_0}\xi_0 + \ln\xi_0 \qquad (5.17)$$

where the value calculated from Eq. (5.10) is substituted for ξ_0.

(b) Analysis employing the mean creep value

As in Sect. 3.2, the calculation of the position of the neutral axis for sections with cracks can utilize the simplification which is provided by using the mean value of creep. If the strain from Eq. (3.54) is introduced into Eq. (5.11) then for a constant modulus of elasticity E_{c0}

$$\frac{\sigma_c(t)}{E_{c0}} + \frac{\sigma_{c0} + \sigma_c(t)}{2E_{c0}}\varphi(t) = \frac{\sigma_s(t)}{E_s}\frac{x(t)}{d_0 - x(t)}$$

Here, σ_{c0} is the initial stress in concrete, i.e. the stress corresponding to the position x_0 of the neutral axis. If the expressions for $\sigma_c(t)$ and $\sigma_s(t)$ from Eq. (5.14) are introduced and, for σ_{c0}, an expression similar to that for $\sigma_c(t)$ is substituted, i.e.

$$\sigma_{c0} = \frac{M}{bx_0\left(d_0 - \dfrac{x_0}{2}\right)}$$

then, at $x(t) = \xi(t)\,d_0$, we can write

$$\frac{1}{(2 - \xi_0)\xi_0}\frac{\varphi(t)}{2}\xi^3(t) - \left(\frac{1}{n_0\mu} + \frac{3}{(2 - \xi_0)\xi_0}\frac{\varphi(t)}{2}\right)\xi^2(t) -$$

$$- \left(1 + \frac{\varphi(t)}{2} - \frac{1}{(2 - \xi_0)\xi_0}\varphi(t)\right)\xi(t) + 1 + \frac{\varphi(t)}{2} = 0 \qquad (5.18)$$

ξ_0 determines the position of the neutral axis for $t = 0$ and is calculated from Eq. (5.10).

(c) Approximate analysis based on the variation of the modulus of deformability

Equations (5.17) and (5.18) are inconvenient for practical use, because the calculation of the position of the neutral axis $x(t)$ is difficult. For this reason, the analysis is simplified by using the modulus of deformability which depends on time through the creep coefficient (see Eq. (5.3b)), viz.

$$E_c(t) = 0.5\frac{1}{1 + \varphi(t)}E_c = \frac{1}{1 + \varphi(t)}E_{c0}$$

The position of the neutral axis is then calculated from Eq. (5.10), i.e.

$$\xi(t) = \frac{n(t)\,\mu}{2}\left(-1 + \sqrt{1 + \frac{4}{n(t)\,\mu}}\right) \tag{5.19}$$

where $n(t) = E_s/E_c(t)$.

Example on the analysis of the position of the neutral axis

The differences in the position of the neutral axis resulting from the individual methods of analysis are demonstrated by an example of the simply reinforced rectangular cross-section shown in Fig. 5.13. The cross-sectional area of steel $A_s = 27.12\ \text{cm}^2$, $E_s = 0.21 \times 10^6\ \text{MPa}$, $E_c = 0.030\,5 \times 10^6\ \text{MPa}$, $n = 0.21/0.030\,5 = 6.89$, $n_0 = 0.21/0.5 \times 0.030\,5 = 13.8$, $\mu = 27.14/30 \times 56.8 = 0.015\,9$. The creep coefficient is calculated with reference to Eq. (5.4), i.e.

$$\varphi = \beta\frac{\delta}{\vartheta}$$

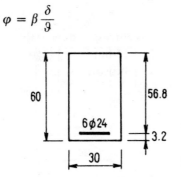

6 φ 24

60   56.8

3.2

30

Fig. 5.13 Reinforced concrete section

The cross-section, for which the neutral axis is required, subjected to a load at time $t_0 = 28\ \text{days} = 0.077$ of a year (i.e. $\vartheta = 1$) and $x(t)$, is to be determined at time $t \to \infty$; the member is stored under dry ambient conditions, i.e. $\delta = 2.85$. Then $\varphi(t) = \beta\delta = 2.85 \times 1.2\,e^{-1.6 \times 0.077} = 3.02$.

At the time when the member was just subjected to the load, i.e. at instant $t = 0$ (considered from time t_0), the following applies (see Eq. (5.10)):

$$\xi_0 = \frac{13.8 \times 0.015\,9}{2}\left(-1 + \sqrt{1 + \frac{4}{13.8 \times 0.015\,9}}\right) = 0.371$$

(a) More accurate analysis

The constant A is calculated from Eq. (5.17), whose numerical form and value are

$$A = \left(\frac{4}{13.8 \times 0.015\,9} + 1\right)\ln(2 - 0.371) +$$

$$+ \frac{1}{13.8 \times 0.015\,9}\left(\frac{1}{1 - 0.371} + 0.371\right) + \ln 0.371 = 17.328\,4$$

236

Table 5.3 Tentative determination of the position of the neutral axis

$\xi(t)$	$a_1 \ln [2-\xi(t)]$	$a_2 \dfrac{1}{1-\xi(t)}$	$a_2\xi(t)$	$\ln \xi(t)$	$\varphi(t)$
0.4	9.038 1	7.595 8	1.823 0	−0.916 3	0.212 2
0.5	7.797 1	9.115 0	2.278 8	−0.693 1	1.169 4
0.6	6.470 3	11.393 8	2.734 5	−0.510 8	2.759 4
0.61	6.332 5	11.685 9	2.780 1	−0.494 3	2.975 8
0.613	6.290 9	11.776 5	2.793 7	−0.489 4	3.04 ≑ 3.02

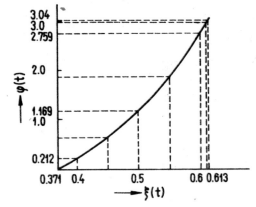

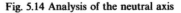

Fig. 5.14 Analysis of the neutral axis

The constants of the terms containing the unknown $\xi(t)$ in Eq. (5.16) are

$$a_1 = \frac{4}{13.8 \times 0.015\,9} + 1 = 19.229\,9; \quad a_2 = \frac{1}{13.8 \times 0.015\,9} = 4.557\,5$$

The most convenient method of solving Eq. (5.16) is to assume several values of $\xi(t)$, corresponding to the actual position of the neutral axis, so that $\varphi(t)$ is calculated; the values which are close to the given $\varphi(t)$ (i.e. 3.02 in this case) are used and then $\xi(t)$ is determined by means of interpolation between the neighbouring values. The procedure is expressed numerically in Table 5.3 for the entire distribution of creep from time t_1 until $t_2 \to \infty$; the results are shown in Fig. 5.14, where the displacement of the neutral axis may be observed.

The position of the neutral axis is

$$\xi(t) = 0.613$$

Note:

$$\xi(t) = \xi_0 \quad \text{for } \varphi(t) = 0.$$

237

(b) Analysis with the aid of the mean creep value

The coefficients of $\xi(t)$ in Eq. (5.18) are

$$\frac{1}{(2 - 0.371)\,0.371} \times \frac{3.02}{2} = 2.499$$

$$\frac{1}{13.8 \times 0.0159} + \frac{1}{(2 - 0.371)\,0.371} \times \frac{3.02}{2} = 12.053$$

$$1 + \frac{3.02}{2} + \frac{3.02}{(2 - 0.371)\,0.371} = -2.487$$

$$1 + \frac{3.02}{2} = 2.501$$

and, after substitution, the equation becomes

$$2.499\xi^3(t) - 12.053\xi^2(t) + 2.487\xi(t) + 2.501 = 0$$

The solution yields

$$\xi(t) = 0.621\,5$$

(c) Approximate analysis

The modulus of deformability

$$E_c(t) = 0.5\,\frac{1}{(1 + 3.02)} \times 0.030\,5 \times 10^6 = 0.003\,794 \times 10^6$$

and

$$n(t) = \frac{0.21}{0.003\,794} = 55.36$$

so that

$$\xi(t) = \frac{55.36 \times 0.015\,9}{2} \times \left(-1 + 1 + \sqrt{\frac{4}{55.36 \times 0.015\,9}}\right) = 0.596$$

It is evident from the comparison of the three values of $\xi(t)$ that the approximate analysis yields a quite satisfactory result; compared with the "accurate" analysis, $\xi(t)$ obtained with the aid of the mean value is 1 per cent higher, and that determined by means of the approximate method is 3 per cent lower. Considering the inaccuracies which encumber the creep coefficient, these differences are very small. It should be noted that, in other cases (for example, with another steel ratio), the differences may be somewhat higher; in spite of this, the approximate analysis can be recommended as a reliable method.

238

The position $x(t)$ of the neutral axis (i.e. with regard to the effects of creep) is strongly influenced by the reinforcement of the cross-section under compression, because the compressed reinforcing bars reduce the creep of concrete.

The analysis of doubly reinforced cross-sections by the accurate method (a) is possible, when the initial assumptions (1) to (5) are complemented by a further assumption that the stress in the compressed reinforcement is

$$\sigma_c(t) \frac{E_s}{E_c(t)}$$

Note: In this case, the compressed reinforcement should be a maximum of 0.5 of the distance from the outermost compressed fibres of the concrete. The position of the neutral axis in a rectangular doubly reinforced cross-section is then calculated from the following differential equation:

$$\frac{1}{n\mu} \frac{-0.5\xi^4(t) - a(t)\,\xi^3(t) + b(t)\,\xi^2(t) + c(t)\,\xi + d(t)}{-0.5\xi^4(t) + 2\xi^3(t) - e(t)\,\xi^2(t) + f(t)\,\xi + g(t)}\, d\xi = d\varphi$$

with

$$a(t) = (1 + \varphi(t))\,(\mu' n_0 + \mu n_0)$$

$$b(t) = 1 + (1 + \varphi(t))\,[\mu' n_0(0.5 + \alpha') + 3\mu n_0]$$

$$c(t) = 2\mu n_0(1 + \varphi(t))\,(1 + \alpha') - 3(1 + \varphi(t))\,\mu n_0$$

$$d(t) = \mu'^2 n_0^2(1 + \varphi(t))^2\,(1 - \alpha') + (1 + \varphi(t))\,\mu n_0$$

$$e(t) = 2.5 - g(t); \quad f(t) = 1 - 2g(t); \quad g(t) = \mu' n_0(1 + \varphi(t))\,(1 - \alpha')$$

$$n_0 = \frac{E_s}{E_{c0}} = 2\frac{E_s}{E_c}; \quad \alpha' = \frac{a'}{d_0}$$

where a' is the distance of the centre of action of the forces in the compressed steel from the outermost fibres of the compressed edge of the concrete. Also,

$$\mu = \frac{A_s}{bd_0}; \quad \mu' = \frac{A_s'}{bd_0}$$

The introduced differential equation can only be solved graphically and is too complex for practical use. Hence, the position of the neutral axis in doubly reinforced sections is calculated approximately [5.3] in the same way as if it were a simply reinforced section. In this case, the reduced creep coefficient (see Eq. (3.10)) is used, i.e.

$$\varphi'(t) = \frac{1}{\mu_x n_0}\,(1 - e^{-\alpha_0\varphi}) \tag{5.20}$$

where

$$\alpha_0 = \frac{\mu_x n_0}{\mu_x n_0 + 1}$$

Here, $\mu_x = A'_s/A_{cc}$ in which A'_s is the area of the compressed reinforcement and A_{cc} is the area of the cross-section under compression. The deformation of the compressed region varies linearly from its highest value at the furthest compressed fibre from the neutral axis down to zero at the neutral axis and, therefore, only the compressed reinforcement located at a sufficient distance from the neutral axis can be taken into account. According to some recommendations, this is only the compressed reinforcement whose distance from the compressed edge does not exceed half the distance of the neutral axis from the same edge.

Then the coefficient $\varphi'(t)$ can be actually determined only by approximation, because the position $\xi(t)$, which is a function of the coefficient $\varphi'(t)$, is not known in advance; considering, however, the approximate character of the analysis with regard to the effect of the compressed reinforcement on the position of the neutral axis, it seems reasonable to calculate the compressed part of concrete using the position ξ_0, as determined for a doubly reinforced cross-section. In this case, the modulus of elasticity will be E_{co} $(= 0.5E_c)$.

The same approach may be adopted for the introduction of the effect of compressed reinforcement, when the method using the mean value of creep is applied.

The approximate analysis of method (c) assumes that the stress in the compressed reinforcement, whose location must satisfy the earlier mentioned condition, is

$$\sigma_c \frac{E_s}{E_c(t)} = \sigma_c \frac{E_s}{0.5 \dfrac{1}{1 + \varphi(t)} E_c}$$

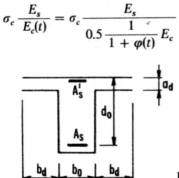

Fig. 5.15 Reinforced concrete *T*-section

If the initial assumptions given in paragraphs (1) to (4) are complemented by this assumption, $\xi(t)$ may be calculated from the conditions of deformation of the cross-section and from the conditions of equilibrium. The resulting expressions below apply for a *T*-section (Fig. 5.15).

$$\xi(t) d_0 < d_d$$

$$\xi(t) = -\frac{n(t)(\mu' + \mu)}{2(\bar{\beta} + 1)} + \sqrt{\left(\frac{n(t)(\mu' + \mu)}{2(\bar{\beta} + 1)}\right)^2 + \frac{n(t)\mu}{\bar{\beta} + 1}} \leq \bar{\alpha} \qquad (5.21a)$$

$$\xi(t) d_0 > d_d$$

240

$$\xi(t) = -\frac{n(t)(\mu' + \mu) + \bar{\alpha}\bar{\beta}}{2} + \sqrt{\left(\frac{n(t)(\mu' + \mu) + \bar{\alpha}\bar{\beta}}{2}\right)^2 + n_0\mu} \geq \bar{\alpha} \qquad (5.21b)$$

Here

$$\bar{\alpha} = \frac{d_d}{d_0}; \qquad \bar{\beta} = \frac{2b_d}{b_0}$$

The expression for a rectangular section is derived from Eq. (5.21a) with $\bar{\beta} = 0$

$$\xi(t) = -\frac{n(t)(\mu' + \mu)}{2} + \sqrt{\left(\frac{n(t)(\mu' + \mu)}{2}\right)^2 + n(t)\mu} \qquad (5.21c)$$

$$\mu' = \frac{A'_s}{b_0 d_0}; \qquad \mu = \frac{A_s}{b_0 d_0}$$

Example on the analysis of the position of the neutral axis of a doubly reinforced cross-section

The analysis will be explained for the member described in the foregoing example; besides the tensile reinforcement, the cross-section is reinforced in its compressed part (Fig. 5.16). Other parameters, such as the moduli of elasticity, the tensile reinforcement and the age of the member at loading and checking of the member, are the same as those of the simply reinforced cross-section.

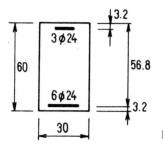

Fig. 5.16 Reinforced concrete rectangular section

The area of the compressed reinforcement equals half of the area of the tensile reinforcement, i.e. $A'_s = 13.57 \text{ cm}^2$, and its axis is 3.2 cm distant from the compressed edge of the cross-section.

"Accurate" analysis

$$\mu' = \frac{13.57}{30 \times 56.8} = 0.008\,0$$

The position of the neutral axis at the instant of loading is calculated from Eq. (5.21c) in which n_0 is substituted for $n(t)$; then, we may write

$$\xi_0 = -\frac{13.8(0.008 + 0.015\,9)}{2} +$$

$$+ \sqrt{\left(\frac{13.8(0.008 + 0.015\,9)}{2}\right)^2 + 13.8 \times 0.015\,9} = 0.332$$

241

$x_0 = 18.86$ cm, and hence $0.5x_0 > 3.2$ cm so that the compressed reinforcement can be considered in the analysis as being fully effective.

$$\mu_x = \frac{13.57}{30 \times 0.332 \times 56.8} = 0.024$$

$$\alpha_0 = \frac{0.024 \times 13.8}{0.024 \times 13.8 + 1.0} = 0.248\,8$$

$$\varphi'(t) = \frac{1}{0.024 \times 13.8}(1 - e^{-0.2488 \times 3.02}) = 1.595$$

To complete the analysis, let us consider also the effect of the reinforcement on the modulus of deformability of that part of concrete under compression by employing the fictitious modulus in accordance with

$$E_c' = E_c(1 + n\mu) \qquad \text{(see Sect. 3.2)}$$

$$E_{c0}' = 0.5 \times 0.050\,5 \times 10^6(1 + 0.024 \times 6.89) = 0.017\,6 \times 10^6$$

into which the value calculated by means of the modulus of elasticity E_c has been substituted for n.

$$n_0 = \frac{0.21}{0.017\,8} = 11.8$$

The position of the neutral axis in an equivalent simply reinforced cross-section is determined by the coefficient

$$\xi_0 = \frac{11.8 \times 0.015\,9}{2}\left(-1 + \sqrt{1 + \frac{4}{11.8 \times 0.015\,9}}\right) = 0.349$$

$$A = \left(\frac{4}{11.8 \times 0.0159} + 1\right)\ln(2 - 0.349) + \frac{1}{11.8 \times 0.0159} \times$$

$$\times \left(\frac{1}{1 - 0.349} + 0.349\right) + \ln 0.349 = 20.1854$$

The equation for the calculation of $\xi(t)$, in accordance with Eq. (5.16), is

$$22.319\,7 \ln(2 - \xi(t)) + 5.329\,9\left(\frac{1}{1 - \xi(t)} + \xi(t)\right) +$$

$$+ \ln \xi(t) - 20.1854 = \varphi'(t)$$

yielding

$$\xi(t) = 0.508$$

242

Analysis with the aid of the mean creep value

In this case, the basic Eq. (5.18) used for the analysis has the form

$$1.3841\zeta^3(t) - 9.4821\zeta^2(t) + 0.9706\zeta(t) + 1.7975 = 0$$

yielding

$$\zeta(t) = 0.511$$

Approximate analysis

Equation (5.21c), after substituting numerical values, gives

$$\zeta(t) = \frac{55.35(0.008 + 0.015\,9)}{2} +$$

$$+ \sqrt{\left(\frac{55.35(0.008 + 0.015\,9)}{2}\right)^2 + 55.35 \times 0.015\,9} = 0.486$$

Here, the ratio $n(t)$ has been calculated from the final value of the creep coefficient, i.e. $\varphi(t) = 3.02$ (see the preceding example).

Here again, a comparison of the results shows that the position of the neutral axis determined by means of the approximate analysis is quite acceptable for practical use, because the values of $\zeta(t)$ determined with the aid of either method only differ by approximately 4 per cent.

The examples of rectangular cross-sections with simple and double reinforcement indicate that the compressed part of concrete is greatly increased due to the long-term effect of loading: this increase represents 65 per cent for simply reinforced and 53 per cent for doubly reinforced cross-sections. Sometimes, this may significantly influence the magnitude of the rigidity. Hence, it is frequently useful to consider the effect of creep on the position of the neutral axis, because the theoretical rigidity of the cross-section increases.

Note: In spite of the given examples in which the position $x(t)$ is analysed, it should be noted that the assumptions representing the basis of these analyses do not express all factors. Thus, for example, the position $x(t)$ is also influenced by the magnitude of the bending moment M_a of the extermal forces, a fact which has been considered to some degree in the formula ascertained from the basis of experiments. The formula has been included in the Soviet Standard and for members subjected to bending:

$$\zeta = \frac{nA_s}{1.8nA_s + 0.1b_0d_0 + \dfrac{1}{4}\left[2(A'_p + n_0A'_s)\dfrac{z_p}{d_0} + \dfrac{M_a}{d_0R_{cb}}\right]}$$

where A'_p for a *T*-section is the area of the outstanding regions of the compressed flange (slab);

z_p is the distance of the centroid, determined for the area A'_p and for the cross-sectional area of the compressive reinforcement $(n_0 A'_s)$ from the centroid of the tensile reinforcement.

For a rectangular cross-section, with reference to Fig. 5.13

when $M_a = 0$ $\xi = 0.369$
when $M_a = 78.4\,\text{kNm}$ $\xi = 0.353$
when $M_a = 158.7\,\text{kNm}$ $\xi = 0.339$

In the first case, when $M_a = 0$, the result agrees satisfactorily with that yielded by Eq. (5.10) (refer to the example with $\xi = 0.371$), while in the other two cases the above formula leads to a lower rigidity of the cross-section. This applies also to the position $x(t)$ under long-term loading.

5.1.2.2 Structural members subjected to eccentric compression and tension

For cross-sections stressed in this way, the "accurate" analysis is very complex, and hence, it is useful to use the approximate analysis which is justified by the conclusions stated earlier. The position of the neutral axis is determined by the normal procedure: from the conditions of equilibrium and from the condition of deformation of the cross-section. Subsequently, the derivation of the expressions for the eccentric compression and the resulting formula for the eccentric tension are given.

Eccentric compression

The conditions of equilibrium can be written, with reference to Fig. 5.17

$$N_c + N'_s - N_s - N = 0$$
$$N_c z_c + N'_s z'_s + N_s z_s - N(x(t) \pm \bar{e}) = 0$$

The external force acting at a distance $\bar{e}$ from the compressed edge of the cross-section, is designated N. In order to reduce the notation, only the positions of the neutral axis $x(t)$ and $n(t)$ are designated as functions of time, although the internal forces, their lever-arms and the stresses depend on time as well.

Assuming a linear relationship between the strain of a fibre and its distance from the neutral axis, the above conditions can be written in the form

$$\int_0^{x(t)} \sigma_c b_x \, dx + n(t)\, \sigma_c \frac{x(t) - a'}{x(t)} A'_s - n(t)\, \sigma_c \frac{d_0 - x(t)}{x(t)} A_s - N = 0$$

$$\int_0^{x(t)} \sigma_c b_x \, dx + n(t)\, \sigma_c \frac{(x(t) - a')^2}{x(t)} A'_s + n(t)\, \sigma_c \frac{(d_0 - x(t))^2}{x(t)} A_s - N(x(t) \pm \bar{e}) = 0$$

244

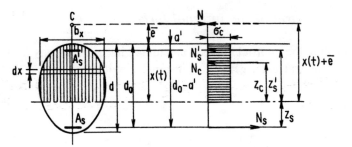

Fig. 5.17 Internal forces with eccentric compression

Designating $\int_0^{x(t)} b_x \, dx = A_x$ and $\int_0^{x(t)} x b_x \, dx = S_x$, we can write

$$\frac{\sigma_c}{x(t)} \left[A_x x(t) + n(t) A_s'(x(t) - a') - n(t) A_s(d_0 - x(t)) \right] - N = 0$$

$$\frac{\sigma_c}{x(t)} \left[S_x x(t) + n(t) A_s'(x(t) - a')^2 + n(t) A_s(d_0 - x(t))^2 \right] - N(x(t) \pm \bar{e}) = 0$$

and, after elimination of N, the position $x(t)$ is given by the relation

$$x(t) \pm \bar{e} = \frac{S_x x(t) + n(t) A_s'(x(t) - a')^2 + n(t) A_s(d_0 - x(t))^2}{A_x x(t) + n(t) A_s'(x(t) - a') - n(t) A_s(d_0 - x(t))^2} \tag{5.22}$$

In the case of the more common rectangular section, Eq. (5.22) is

$$x(t) \pm \bar{e} = \frac{0.5bx^3(t) + n(t) A_s'(x(t) - a')^2 + n(t) A_s(d_0 - x(t))^2}{bx^2(t) + n(t) A_s'(x(t) - a') - n(t) A_s(d_0 - x(t))}$$

Eccentric tension

Using the designation of Fig. 5.18 and proceeding as before, the following expression is obtained

$$x(t) - \bar{e} = \frac{S_x x(t) + n(t) A_s'(x(t) - a')^2 + n(t) A_s(d_0 - x(t))^2}{A_x x(t) + n(t) A_s'(x(t) - a') - n(t) A_s(d_0 - x(t))}$$

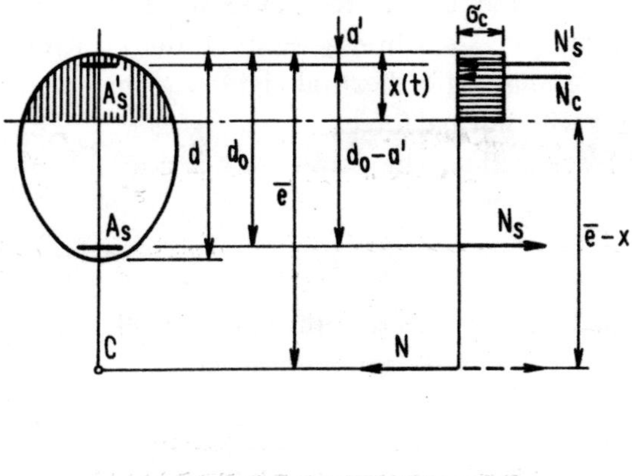

Fig. 5.18 Internal forces with eccentric tension

5.1.2.3 *The lever-arm of the internal forces*

The lever-arm of the internal forces is calculated in the usual way from the known position of the neutral axis; generally, it can be written

$$z(t) = \frac{S_n(t)}{A_n(t)} \tag{5.23}$$

where $S_n(t)$ is the static moment of the equivalent area $A_n(t)$ of the compressed part of the cross-section about the axis passing through the centroid of the tensile reinforcement; the equivalent area is calculated from the relation

$$A_n(t) = A_x(t) + n(t)\, A'_s$$

where $A_x(t)$ is the cross-sectional area of compressed concrete, A'_s is the area of the compressed reinforcement and $n(t) = E_s/E_c(t)$.

5.1.3 Rigidity of cross-section

5.1.3.1 *Effect of creep*

When the redistribution of stresses induced by creep is analysed, and when the strain due to long-term loading is calculated for reinforced-concrete structural members in which cracks have developed in the service state, the flexural rigidity is determined by means of the average strains of the member. The average curvature between the cracks is determined from the expression (Fig. 5.19)

$$\frac{1}{\varrho} = \frac{\varepsilon'_s(t) + \varepsilon'_c(t)}{d_0} \tag{5.24}$$

where $\varepsilon'_s(t)$ and $\varepsilon'_c(t)$ are the average strains of steel and concrete, respectively, in the section between cracks (Fig. 5.6). This approach is quite analogous to the principle, which applies in the calculations of the curvature of a member subject to short-term loading; consequently, the procedure for the derivation of the expressions needed for the calculation of the curvature is the same and the resulting form of the expression is similar. We shall briefly describe the principle of derivation of the formula used for the analysis of flexural rigidity of members subjected to pure bending and to bending with normal forces.

The average strains $\varepsilon'_s(t)$ and $\varepsilon'_c(t)$ are defined in Sects. 5.1.1.2 and 5.1.1.3, where it has been demonstrated that

$$\varepsilon'_s(t) = \psi_s(t)\, \varepsilon_s \qquad\qquad \varepsilon'_c(t) = \psi_c(t)\, \varepsilon_c$$

Here, ε_s is the average strain of tensioned steel at a section through a crack and ε_c is the average strain of compressed concrete at a section without crack.

246

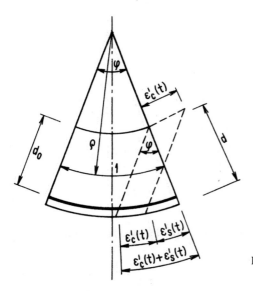

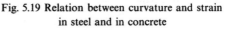

Fig. 5.19 Relation between curvature and strain
in steel and in concrete

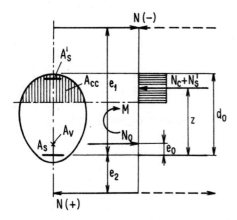

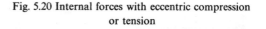

Fig. 5.20 Internal forces with eccentric compression
or tension

Consider a structural member subjected to external forces of moment M and to a compressive $(-N)$ or tensile $(+N)$ force at a distance e_1 or e_2 from the axis of the tensile reinforcement; the member is reinforced in both the tensioned and compressed parts of the cross-section (Fig. 5.20). If the member is also prestressed by a force N_0, which acts at a distance e_0, then the quantities M_t (total moment) and N_t (total force) acting on the section are

$$M_t = M + Ne_1 + N_0e_0 \quad (\text{or } M_t = M + Ne_2 + N_0e_0)$$
$$N_t = N - N_0$$

It should be noted, with regard to the prestressing force N_0, that it should be such a value as to allow the formation of cracks in the tensioned concrete.

247

The stresses in concrete and steel are

$$\sigma_c(t) = \frac{M_t}{z(t) \, A_n(t)}$$

$$\sigma_s(t) = \frac{1}{A_s} \left(\frac{M_t}{z(t)} + N_t \right)$$

while the average strains of concrete and steel are

$$\varepsilon'_c(t) = \frac{\psi_c(t) \, M_t}{z(t) \, E_c(t) \, A_n(t)}$$

$$\varepsilon'_s(t) = \frac{\psi_s(t)}{E_s A_s} \left(\frac{M_t}{z(t)} + N_t \right)$$

When both these expressions are introduced into Eq. (5.24), the formula for the calculation of the curvature is

$$\frac{1}{\varrho} = \frac{M_t}{d_0 z(t)} \left(\frac{\psi_s(t)}{E_s A_s} + \frac{\psi_c(t)}{E_c(t) \, A_n(t)} \right) + \frac{N_t}{d_0} \frac{\psi_s(t)}{E_s A_s} \tag{5.25}$$

in which the meaning of the symbols $z(t)$ and $A_n(t)$ has been explained in Sect. 5.1.2.3.

Considering that

$$\frac{1}{\varrho} = \frac{M_t}{B(t)}$$

where $B(t)$ is the flexural rigidity, for members subjected to pure bending

$$B(t) = \frac{d_0 z(t)}{\dfrac{\psi_s(t)}{E_s A_s} + \dfrac{\psi_c(t)}{E_c(t) \, A_n(t)}} \tag{5.26}$$

Example on the analysis of flexural rigidity

Let us analyse the flexural rigidity of a rectangular cross-section with double reinforcement (Fig. 5.16)

$$\psi_s = 1.0 \qquad \psi_c = 0.9$$
$$x(t) = 0.508 \times 56.8 = 28.9 \text{ cm}$$
$$A_n(t) = 28.9 \times 30 + 55.35 \times 13.57 = 1\,618 \text{ cm}^2$$
$$S_n(t) = 28.9 \times 30 \, (56.8 - 0.5 \times 28.9) + 55.35 \times 13.57 \times$$
$$\times \, (56.8 - 3.2) = 76\,976 \text{ cm}^3$$

The lever-arm of the internal forces, in accordance with Eq. (5.23), is

$$z(t) = \frac{76\,976}{1\,618} = 47.6 \text{ cm}$$

and the flexural rigidity is

$$B(t) = \cfrac{0.568 \times 0.476}{\cfrac{1}{0.21 \times 10^6 \times 0.002\,714} + \cfrac{0.9}{0.003\,794 \times 10^6 \times 0.161\,8}} =$$

$$= 83.9 \text{ MN m}^2$$

5.1.3.2 *Effect of shrinkage*

Assuming equal conditions for the distribution of shrinkage over the entire section, the fibres of a reinforced member are shortened due to shrinkage so that the strain (ε_{cs}) at the edge without reinforcement is larger than that at the reinforced edge (ε'_{cs}), because the reinforcement restrains the free development of the deformation (Fig. 5.21). If the strain is related to the deformation of the reinforcement, the curvature will be

$$\frac{1}{\varrho} = \frac{\varepsilon_{cs} - \varepsilon_{ss}}{d_0} = \frac{\varepsilon_{cs}}{d_0}\left(1 - \frac{\varepsilon_{ss}}{\varepsilon_{cs}}\right)$$

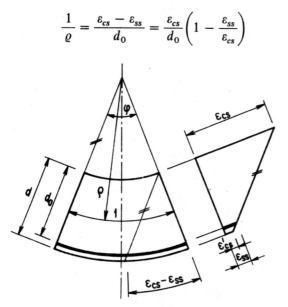

Fig. 5.21 Curvature caused by shrinkage

Here, ε_{cs} designates the shrinkage of concrete which would develop without any restraint, and ε_{ss} is the strain of the reinforcing bars; d_0 is the effective depth of the cross-section. Cracks form in the tensioned concrete, and their effect on the strain of steel can be found only experimentally by determining the value of the ratio $\varepsilon_{ss}/\varepsilon_{cs}$. This value, given in references [5.2, 5.5], is equal to

    0.3   for members with a low steel ratio;

    0.1   for members with a high steel ratio.

A more accurate formula, taking into account the steel ratios of the compressive and tensile reinforcement, is

$$\frac{1}{\varrho} = 0.7 \frac{\varepsilon_{cs}}{d} (\mu - \mu')^{\frac{1}{3}} \left[\frac{\mu - \mu'}{\mu} \right]^{\frac{1}{2}} \qquad \text{if } \mu - \mu' \leq 3\% \qquad (5.27a)$$

$$\frac{1}{\varrho} = \frac{\varepsilon_{cs}}{d} \qquad \text{if } \mu - \mu' > 3\% \qquad (5.27b)$$

In these formulae, μ and μ' are in per cent and they relate to the entire depth of the cross-section.

Example of analysis

The analysis of curvature is related to the example in Fig. 5.16, where the cross-section is doubly reinforced. The shrinkage $\varepsilon_{cs} = 0.000\,22$.

$$\mu = 100 \frac{27.14}{30 \times 60} = 1.50\% \qquad \mu' = 100 \frac{13.57}{30 \times 60} = 0.75\%$$

$$\frac{1}{\varrho} = 0.7 \frac{0.000\,22}{0.60} (1.50 - 0.75)^{\frac{1}{3}} \times \left(\frac{1.50 - 0.75}{1.50} \right)^{\frac{1}{2}} =$$

$$= 1.648 \times 10^4 \text{ m}^{-1}$$

5.1.3.3 *Distribution of rigidity*

The rigidity of cross-sections in a structural member varies in relation to the arrangement and ratio of reinforcement, and it also depends on the cross-section being solid or with cracks. The variation of rigidity is portrayed in Fig. 5.1, where it may be observed that the section acts as a solid unit at the points of small bending moment. In this case, the rigidity is calculated from the formula

$$B(t) = \frac{J_i E_c}{1 + \varphi(t)}$$

where J_i is the moment of inertia of an ideal cross-section, which is calculated as a sum of the entire area of concrete and the n-fold area of reinforcement, n being equal to $\frac{E_s}{E_c}$. Under short-term loading, the rigidity of these solid cross-sections is always higher than that of identical cross-sections with the same reinforcement, but with eliminated tensile stress. Under long-term loading, the area of the compressed part of the cross-section increases in time; consequently, it may happen that the fully acting sections of a beam can sometimes have even a lower rigidity. This disagreement is due to the different assumptions of the analysis of fully acting cross-sections and of those with cracks.

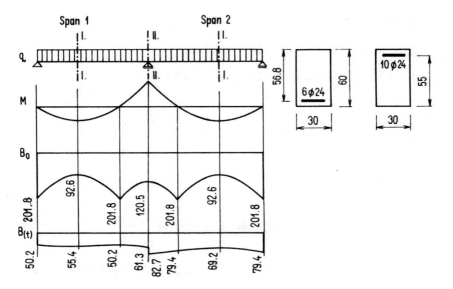

Fig. 5.22 Variation of short- and long-term rigidity

Let us demonstrate the variations of rigidity on the example of a continuous beam of two equal spans. The structure is composed of two parts which were precast and then connected by welding their reinforcing bars and by filling the joint with concrete above the intermediate support when the age of one part was 28 days and that of the other part was 185 days.

The cross-sectional dimensions are 30 by 60 cm and the area of the reinforcement is 27.14 cm^2 at mid-spans and 45.23 cm^2 at the support. The rigidities of the cross-sections are to be determined for $t \to \infty$.

The creep coefficient is (with reference to Sect. 5.1.1) ($\vartheta = 1.0$; $\delta = 2.85$)

for the first span: $\varphi(t) = 2.85 \times 1.2\,e^{-1.6 \times 0.077} = 3.02$
$(t_0 = 28 \text{ days} = 0.077 \text{ of a year})$

for the second span: $\varphi(t) = 2.85 \times 1.2\,e^{-1.6 \times 0.5} = 1.54$
$(t_0 = 185 \text{ days} = 0.5 \text{ of a year})$

The modulus of elasticity of concrete $E_c = 0.0305 \times 10^6$ MPa. Other pertinent data are given in Fig. 5.22.

Let us assume that both spans of the beam are subjected to a uniformly distributed load and that the bending moments are distributed in accordance with Fig. 5.22. The fully acting cross-sections then are located near the freely supported ends of the beam and near the points where the moment changes its sign.

Rigidity under short-term loading

$$n = 6.89; \quad \psi_s = 1.0; \quad \psi_c = 0.9$$

251

At fully acting sections

$$A_i = 30 \times 60 + 6.89 \times 27.14 = 1987 \, \text{cm}^2$$

The distance of the centroid from the cross-sectional edges

$$t_1 = 27.48 \, \text{cm}; \quad t_2 = 32.52 \, \text{cm}$$

$$J_i = \frac{1}{3} 0.30(0.325 \, 2^3 + 0.274 \, 8^3) + 6.89 \times 27.14(0.274 \, 8 - 0.032)^2 =$$

$$= 0.006 \, 616 \, \text{m}^4$$

$$B_0 = 0.0305 \times 10^6 \times 0.006 \, 616 = 201.8 \, \text{MN m}^2$$

$$\mu = 0.0159$$

At the section with the highest positive bending moment

$$\xi_0 = \frac{6.89 \times 0.0159}{2} \times \left(-1 + \sqrt{1 + \frac{4}{6.89 \times 0.0159}} \right) = 0.281$$

$$z_0 = 0.568 \, (1 - 0.5 \times 0.281) = 0.488 \, \text{m}$$

$$B_0 = \frac{0.568 \times 0.488}{\dfrac{1}{0.21 \times 10^6 \times 0.002\,714} + \dfrac{0.9}{0.5 \times 0.0305 \times 10^6 \times 0.3 \times 0.281 \times 0.568}} =$$

$$= 92.6 \, \text{MN m}^2$$

At the support

$$\xi_0 = \frac{6.89 \times 0.0274}{2} \left(-1 + \sqrt{1 + \frac{4}{3.89 \times 0.0274}} \right) = 0.350$$

$$z_0 = 0.55 \, (1 - 0.5 \times 0.350) = 0.454 \, \text{m}$$

$$B_0 = \frac{0.55 \times 0.454}{\dfrac{1}{0.21 \times 10^6 \times 0.004\,523} + \dfrac{0.9}{0.5 \times 0.0305 \times 10^6 \times 0.3 \times 0.35 \times 0.55}} =$$

$$= 120.5 \, \text{MN m}^2$$

Rigidity under long-term loading

$$\psi_s = 1.0; \quad \psi_c = 0.9$$

For the first span

At fully acting sections
The cross-sectional quantities are the same as in the case of short-term loading

$$B(t) = 0.006 \, 616 \frac{1}{1 + 3.02} \times 0.0305 \times 10^6 = 50.2 \, \text{MN m}^2$$

At the section with the highest positive bending moment (with zero tension)

μ $= 0.0159$

$\xi(t) = 0.596$ (from the example to Fig. 5.13)

$z(t) = 0.568\,(1 - 0.5 \times 0.596) = 0.399$ m

$E_c(t) = 0.5\,\dfrac{1}{1 + 3.02} \times 0.030\,5 \times 10^6 = 0.003\,794 \times 10^6$ MPa

$$B(t) = \dfrac{0.568 \times 0.399}{\dfrac{1}{0.21 \times 10^6 \times 0.002\,714} + \dfrac{0.9}{0.003\,794 \times 10^6 \times 0.3 \times 0.596 \times 0.568}} =$$

$= 55.4$ MN m^2

At the support (with zero tension)

μ $= 0.0274$; $A_s = 45.23$ cm^2; $n = 55.35$

$\xi(t) = \dfrac{55.35 \times 0.027\,4}{2}\left(-1 + \sqrt{1 + \dfrac{4}{55.35 \times 0.0274}}\right) = 0.688$

$z(t) = 0.55\,(1 - 0.5 \times 0.688) = 0.361$ m

$$B(t) = \dfrac{0.55 \times 0.361}{\dfrac{1}{0.21 \times 10^6 \times 0.004\,523} + \dfrac{0.9}{0.003\,794 \times 10^6 \times 0.3 \times 0.688 \times 0.55}} =$$

$= 63.1$ MN m^2

For the second span

At fully acting sections

The cross-sectional quantities are the same as in the case of short-term loading

$B(t) = 0.006\,616\,\dfrac{1}{1 + 1.54} \times 0.030\,5 \times 10^6 = 79.4$ MN m^2

$E_c(t) = 0.5\,\dfrac{1}{1 + 1.54} \times 0.030\,5 \times 10^6 = 0.006\,004 \times 10^6$ MPa

$n(t) = \dfrac{0.21}{0.006\,004} = 34.98$

$\xi(t) = \dfrac{34.98 \times 0.015\,9}{2} \times \left(-1 + \sqrt{1 + \dfrac{4}{34.98 \times 0.0159}}\right) = 0.518$

$z(t) = 0.568\,(1 - 0.5 \times 0.518) = 0.421$ m

$$B(t) = \frac{0.568 \times 0.421}{\dfrac{1}{0.21 \times 10^6 \times 0.002\,714} + \dfrac{0.9}{0.006\,004 \times 10^6 \times 0.3 \times 0.518 \times 0.568}} =$$

$$= 69.2 \text{ MN m}^2$$

At the support

$$\xi(t) = \frac{34.98 \times 0.0274}{2} \left(-1 + \sqrt{1 + \frac{4}{34.98 \times 0.027\,4}} \right) = 0.611$$

$$z(t) = 055\,(1 - 0.5 \times 0.611) = 0.382 \text{ m}$$

$$B(t) = \frac{0.55 \times 0.382}{\dfrac{1}{0.21 \times 10^6 \times 0.004\,523} + \dfrac{0.9}{0.006\,004 \times 10^6 \times 0.3 \times 0.611 \times 0.55}} =$$

$$= 82.7 \text{ MN m}^2$$

The values of the calculated rigidities are depicted in Fig. 5.22; the distribution between the calculated points is expressed by a quadratic parabola under short-term loading, this being convenient for eventual further analyses (redistribution of stresses). Under long-term effects, the rigidity in the individual spans is quite uniformly distributed and the calculated points can be connected by a straight line. The example demonstrates how the increase of the compressed cross-sectional zone under long-term loading causes the rigidity of the fully acting sections to decrease much more rapidly than the rigidity of the sections with cracks.

5.1.3.4 *Effect of creep on the redistribution of stresses in a structure*

In statically indeterminate structures, the variation of rigidity along the beam influences the magnitude of the bending moments, whose variation causes further variations of rigidity. A more accurate analysis of the bending moments for this situation would be very tedious so that it would not be worth while, especially when the variations of rigidity along the beam are small (as, for example, in the long-term loading example of Fig. 5.22). If, however, it is necessary to undertake such an analysis, the correct method would be iteration with the following steps:

(1) A distribution of bending moments is assumed, which corresponds to the moment distribution considered given by the theory of elasticity. For this distribution, the rigidities are calculated in the characteristic cross-sections (in accordance with the procedure demonstrated in the example of Fig. 5.22); between the values of rigidity calculated for the individual sections, a quadratic parabola or a straight line is inserted. In simple cases, the rigidities between the points of zero bending moments are considered constant, and equal to the minimum rigidity in this part of the beam.

(2) For the calculated rigidities, a new distribution of the bending moments is computed, and then a new distribution of rigidities determined.

(3) The procedure is repeated, until the differences between two following steps are acceptably small.

5.1.3.5 *Analysis of deflections*

The deflection of a structure at any point may be calculated by the usual methods from the known distribution of bending moments and rigidities; for example, using the principle of virtual work, we can write

$$f(x) = \int_0^l \frac{M(x)}{B(x)} \overline{M}(x) \, dx$$

where $M(x)$ is the bending moment at cross-section x induced by the load, and $\overline{M}(x)$ is the bending moment induced by a unit force acting at the point of the required deflection; in accordance with Mohr's theorem, deflection is determined as the bending moment of a beam with a loading diagram representing the ratio $M(x)/B(x)$.

Now, we shall introduce the principles of an analysis which has been included in the draft of the Recommendations [5.9].

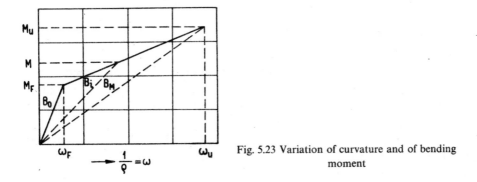

Fig. 5.23 Variation of curvature and of bending moment

The relationship between the bending moment M and the curvature $\omega = \dfrac{1}{\varrho}$ is bi-linear; the first straight line applies for fully acting cross-sections and the other one for cross-sections in which cracks develop with increasing bending moment. If the curvatures at the incipient formation of cracks (for M_F) and at zero tension are known, the respective curvature ω (Fig. 5.23) can be determined for any value of the bending moment $M \, (< M_u)$. For the fully acting moment,

$$\omega_F = \frac{M_F}{E_c J_i} = \frac{M_F}{B_0}$$

255

where E_c is the modulus of elasticity of concrete and J_i is the moment of inertia of the cross-sections (steel reinforcement may be included). At the ultimate moment (M_u), the curvature is

$$(EJ)_U = E_s A_s (d_0 - x_0) z = B_M$$

This expression is derived from the relationship for deformation (Fig. 5.6)

$$\frac{1}{\varrho} = \omega = \frac{\varepsilon_s}{d_0 - x_0} = \frac{\sigma_s}{E_s (d_0 - x_0)}$$

into which $\dfrac{M}{A_s z}$ is substituted for σ_s.

When a cross-section is loaded by a moment $M(M_F < M < M_U)$, the cross-section will have a rigidity B_i with its value between the values of B_0 and B_M (Fig. 5.23); this rigidity can be calculated. To facilitate the calculation of $B_i = E_c J$, the following empirical formula [5.10] is used:

$$J = \left(\frac{M_F}{M_{max}}\right)^3 J_i + \left[1 - \left(\frac{M_F}{M_{max}}\right)^3\right] J_U$$

when

$\qquad M_{max} \qquad$ = maximum bending moment on the beam;

$\qquad J_i$ and J_U = moments of inertia of a fully acting section and of a section with crack, respectively.

Under long-term loading, a reduced modulus of elasticity, $E_c(t) = \dfrac{E_c}{1 + \varphi(t)}$ is used in the analysis.

The Recommendations [5.9] are based on a similar principle, where two cases are distinguished in the analysis of rigidity under short-term loading: (a) when the entire cross-section is active $(M \leq M_F)$, and (b) when cracks develop in the cross-section $(M > M_F)$.

(1) If the beam of a uniform cross-section is subjected to pure bending, its deflection can be calculated only under the assumption that cracks develop in the cross-section, i.e.

$$B = E_s A_s (d_0 - x_0) z$$

However, the reinforcement ratio μ must be at least 0.005 for rectangular sections, and 0.001 for T-sections, where $b/b_0 \geq 10$ (when $1 < b/b_0 < 10$); the minimum μ is determined by interpolation.

For less reinforced cross-sections of symmetrically loaded beams, the mid-span deflection may be obtained from the relation

$$f = kl^2 \left(\frac{M_I}{E_c J_i} + \frac{4}{3} \frac{M_{II}}{E_s A_s (d_0 - x_0) z}\right) \leq kl^2 \frac{M}{E_s A_s (d_0 - x_0) z}$$

256

where l is the span of the member;

$$M_I = M_F;$$
$$M = \text{mid-span bending moment}, M = M_I + M_{II};$$

k = coefficient by which the mid-span bending moment is multiplied; the value of this coefficient for four common cases is given in Fig. 5.24.

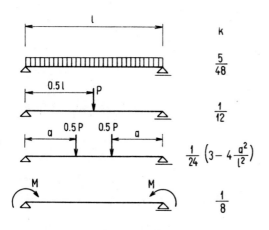

Fig. 5.24 Coefficient for the calculation of deflection

(2) If the beam of a variable cross-section is subjected to a bending moment, or to a normal force, the curvature is calculated with the aid of the following formulae:

Pure bending

at $M \leqq M_F$, $\omega = \dfrac{1}{\varrho} = \dfrac{M}{E_c J_i}$

at $M > M_F$, $\omega = \dfrac{M}{E_s A_s(d_0 - x_0) z}$

Eccentric tension

$$M_I = \frac{M_F}{1 + \dfrac{W_i}{A_i e}}$$

where $e = M/N$ is the eccentricity calculated with respect to the centroidal axis of the cross-section without reinforcement;
W_i and A_i are the sectional modulus and the cross-sectional area of the reinforced section, respectively,

at $M \leqq M_I$, $\omega = \dfrac{M}{E_c J_i}$

257

at $M > M_1,$ $\qquad \omega = \dfrac{M}{E_s A_s(d_0 - x_0)} \left(1 + \dfrac{z - u}{e}\right)$

where u is the distance of the tensile reinforcement from the centroid of the cross-section without reinforcement.

Eccentric compression

$$M_1 = \dfrac{M_F}{1 - \dfrac{W_i}{A_i e}}$$

at $e = M/N \leq d_0 - u,$ $\qquad \omega = \dfrac{M}{E_c J_i}$

at $e > d_0 - u$ and $M \leq M_1,$ $\qquad \omega = \dfrac{M}{E_c J_i}$

at $e > d_0 - u$ and $M > M_1,$ $\qquad \omega = \dfrac{M}{E_s A_s(d_0 - x_0) z} \left(1 + \dfrac{z - u}{e}\right)$

Allowing for the effect of long-term loading in the analysis is very simple. If the loading acts immediately after the completion of the construction, the deformation due to the long-term loading is calculated by multiplying the short-term deformation by a coefficient of 2 for mild or humid ambient conditions; for dry conditions, the coefficient is 3. If the long-term loading starts to act from at least half a year after the attainment of the design strength of concrete, the above coefficients are reduced to 1.5 and 2.0, respectively. The effect of the compressed reinforcement is considered in the analysis by reducing the above coefficients to 60 per cent, when the area of the compressive reinforcement is at least half the area of the tensile reinforcement ($\mu' \geq \mu$); when $\mu' = \mu$, the above coefficients are reduced to 40 per cent.

5.1.4 Effect of creep on buckling combined with bending

The effect of long-term loading on reinforced-concrete columns may be the cause of an increase of the original eccentricity and, in the course of time, of the loss of stability. Consequently, the analysis adopts a different method for determining the effect of creep.

All Standards endeavour to keep formulae as simple as possible to represent actual behaviour. In spite of this, these formulae are still rather complex.

Some methods of analysis, which may be exploited by the use of computers now or in the near future, are introduced in the references [5.11], [5.12], [5.13]; these methods should be used in cases of important structures (bridge piers, etc.).

258

Here, a reasonably accurate analysis will be described, together with the problems connected with these more accurate methods [5.8].

If a column is subjected to a force N with an eccentricity e, the instantaneous deflection will equal y_0. However, the deflection may be so large that the structure loses its stability. Such conditions are given by the relationship between the bending moment M_U of the external forces and the deflection y. To the curve representing this relationship, a straight line P is drawn to represent the relation between the moment M_v of the external forces and the deflection (Fig. 5.25). Thus, for example, the curve K_0 of the relationship $M_0 \sim y_0$, which is at the top of the diagram, relates to the instantaneous loading ($t = 0$), and the straight line P_1 intersects this curve at points B and C; the first point designates a stable state, while the other point designates an unstable one. Another straight line P_2, which is a tangent to the curve K_0, indicates, at a given eccentricity e ($M_v = N_v e$), the critical moment M_{cro} and the deflection y_{cro} by its point of contact D. At these critical values, the structural member becomes unstable and any other straight line above the straight line P lies in the unstable zone. The quantities M_v and N_v are from the external loading.

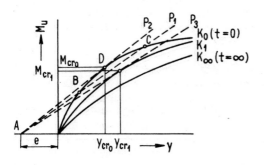
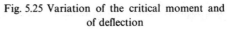

Fig. 5.25 Variation of the critical moment and of deflection

Hence, if the curve K_0 is known, the critical bending moment and the critical deflection for a given eccentricity can be determined by drawing a tangent to this curve and the point of contact gives both these quantities.

Under long-term loading, when time is a factor (or creep coefficient), the curves are situated lower in the diagram and their range is within $\langle K_\infty, K_0 \rangle$. The curve of the relationships M_U versus y depends on the magnitude of the external force, on the relations between the internal forces and strain, on the geometrical characteristics of the member, on the support conditions of the member and also on the time (which affects the internal forces and the strain). The critical moment M_{cr} and the deflection y_{cr}, for the same eccentricity e of the external force N_v, are then different (for example, M_{cr1} and y_{cr1}).

The magnitude of the ultimate strain ε_{cU} varies under the long-term loading; this strain depends on the duration of loading and on the magnitude of the creep coefficient $\varphi(t)$; the distributions shown in Fig. 5.26 were obtained from tests. The

259

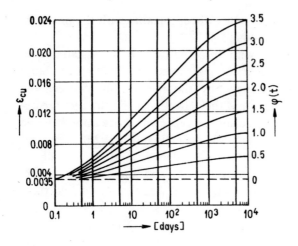

Fig. 5.26 Relation between the ultimate strain ε_{cU} of concrete (in the outermost fibre of a section) and time, and between the coefficient $\varphi(t)$ and time

relationship $M_U \sim y$ is usually determined by using various extreme values of $\varphi(t)$, i.e. after 10 000 days of loading.

If the strain of concrete reaches its ultimate value ε_{cU}, the compressed concrete will be under the effect of the following force:

$$N_{cU} = \alpha_U R_c bx \tag{5.28}$$

where R_c is the strength of concrete;
  $\alpha_U R_c$ is the stress in the compressed part of concrete, assuming a uniformly distributed stress (Fig. 5.27).

The force in concrete acts at a distance of $\beta_U x$ from the compressed edge (Fig. 5.27), while the coefficients α_U and β_U are given in Fig. 5.27 as a function of strength R_c.

If $\varepsilon_c < \varepsilon_{cU}$, the force in the compressed concrete is

$$N_c = k_a \alpha_U R_c bx \tag{5.29}$$

and its point of application is located at a distance of $k_\beta \beta_U x$ from the compressed

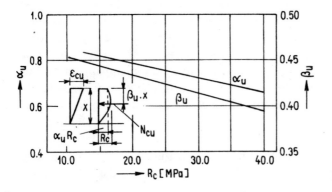

Fig. 5.27 Values of α_U and β_U related to the strength of concrete

edge; the coefficients k_α and k_β are given in Fig. 5.28 for different strengths of concrete and different ratios of $\varepsilon_c/\varepsilon_{cU}$.

Thus, all the values needed for the determination of the behaviour of compressed concrete are known; assuming a linear deformation of the cross-section, equal strain of steel and concrete in the same fibre due to the cohesion of concrete and steel, and elimination of the tensioned part of concrete, the relationship between M_U and y may be found.

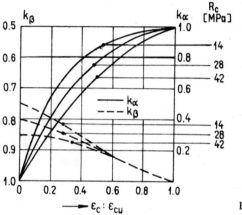

Fig. 5.28 Coefficients k_α and k_β

The determination of any point of the curve representing the relationship between M_U and y may proceed generally as follows:

The following quantities are known: the width b and depth d of the cross-section; the cross-sectional areas of compressive (A'_s) and tensile (A_s) reinforcement; the distance of the point of action of the forces in the reinforcement from the compressed (a) and tensioned (a') edges; the limiting stresses of steel σ_s and σ'_s; the strength of concrete R_c; the magnitude of the external force N and its eccentricity e; the length of the member (l) and the creep coefficient $\varphi(t)$. The ultimate strain ε_{cU} for a given value $\varphi(t)$ is found from Fig. 5.26.

The strain ε_c is chosen within the limits $\langle 0, \varepsilon_{cU} \rangle$ and the coefficients k_α and k_β are found from Fig. 5.28 for the ratio $\varepsilon_c/\varepsilon_{cU}$ and for the strength of concrete R_c. The coefficients α_U and β_U for the given strength of concrete, are found from Fig. 5.27.

To be able to determine the force N_c, which is needed together with the forces N_s and N'_s for the condition of equilibrium, from which the force N is calculated, it is necessary to choose the position x_1 of the neutral axis; then from Eq. (5.29)

$$N_c = k_\alpha \alpha_U R_c b x_1$$

With the known quantities ε_c and x_1, the forces acting in the reinforcement can be calculated, because the strains in the reinforcement are as follows:

261

for the compressed bars

$$\varepsilon'_s = \varepsilon_c \frac{x_1 - d'}{x_1}$$

for the tensioned bars

$$\varepsilon_s = \varepsilon_c \frac{x_1}{d - x_1 - a}$$

The forces acting in the reinforcement are

$$N'_s = \varepsilon'_s E_s A'_s \qquad N_s = \varepsilon_s E_s A_s$$

If, however, the strain $\varepsilon'_s(\varepsilon_s)$ is larger than the ultimate strain of steel, then the forces acting in the reinforcement are calculated from the ultimate stress, viz.

$$N'_s = \sigma_s A'_s \qquad N_s = \sigma_s A_s$$

The accuracy of the estimate of the position of the neutral axis (x_1) is checked by comparing the force $\bar{N}$, resulting from the condition of equilibrium, with the external force N. Thus, the estimate is correct, if both these forces equal, i.e.

$$\bar{N} = N_c + N'_s - N_s \qquad \text{(must equal } N\text{)}$$

If this is not the case, a new force $\bar{N}$ and another position x_2 have to be calculated.

If the agreement between $\bar{N}$ and N is satisfactory, the moment M_U of the internal forces is obtained from the equilibrium on the member, i.e.

$$M_U = N_c(0.5d - k_\beta \beta_U x) + N'_s(0.5d - a') + N_s(0.5d - a)$$

The deflection y can be calculated only if the support conditions at the ends of the member are known; assuming that the member behaves elastically and that its ends are pin-supported, a sine wave may be substituted for the deflection line so that the deflection at section x is

$$y = y_{max} \sin \frac{\pi}{l} x$$

where y_{max} is the deflection at mid-span (for $x = l/2$).

The curvature is then

$$\frac{1}{\varrho} = y'' = -\frac{\pi^2}{l^2} y_{max} \sin \frac{\pi}{l} x$$

To calculate the mid-span deflection, we use

$$\left(\frac{1}{\varrho}\right)_{\frac{l}{2}} = -\frac{\pi^2}{l^2} y_{max}$$

whence,

$$y_{max} = -\left(\frac{1}{\varrho}\right)_{\frac{l}{2}} \frac{l^2}{\pi^2}$$

The curvature of the cross-section is calculated from the strain ε_c and from the position x of the neutral axis, viz.

$$\left(\frac{1}{\varrho}\right)_{\frac{l}{2}} = \frac{\varepsilon_c}{x}$$

hence

$$y_{max} = -\frac{\varepsilon_c}{x} \frac{l^2}{\pi^2}$$

This determines the deflection y related to the moment M_U, i.e. one point of the curve K. Further points are determined in a similar way by chosing a new strain ε_c and repeating the entire procedure until the relationship M_U versus y can be plotted.

The moment induced by the external force N, for $y = 0$, is $M_{v0} = N e$; consequently, the straight line P may be drawn to determine whether the structural member is stable. If this straight line lies above the zone given by the relationship $M_U \sim y$ (curve K), the member is unstable for the given eccentricity and force N; on the other hand, if the straight line intersects the curve, then the member is stable.

6
Effects of shrinkage and creep on structures and measures to be adopted to minimize their unfavourable influence

Experience has shown that structures can be subjected to very large deformations, as well as stresses due to shrinkage and creep; the consequences are excessive deflections of the structural members, wide cracks in the tensioned concrete and redistribution of stresses in statically indeterminate structures. All this greatly affects the behaviour of the material.

These effects have been observed on old concrete structures. However, in the case of new structures, where the strengths of concrete and steel are exploited to a high degree and where the dimensions of the structural members are very slender, creep and shrinkage should be considered with particular care and attention. This Chapter discusses the changes that may occur in the behaviour of residential, office, industrial buildings and bridges, due to shrinkage and creep.

6.1   Residential and office buildings

In the following discussion, two groups of structures are distinguished: frame structures, consisting of one-dimensional members, and panel structures assembled from plate-shaped panels. The distinction is very approximate, because a variety of combinations of both kinds of structure are used in practice.

6.1.1   Frame structures

The effects of both shrinkage and creep have to be taken into account in tall buildings, while in long buildings, the effect of shrinkage is more important than that of creep. Cast-in-situ structures are more sensitive to shrinkage and creep than structures erected from precast members, because in the latter type, creep and shrinkage can develop freely before the prefabricated elements are assembled.

264

Effects in the vertical direction

A concrete frame structure consists of columns and beams, and compatibility of deformations must be maintained in the connections of these members. In any member, a variation of the deformation is followed by a variation of the internal forces. Hence, any shortening of the columns produced by shrinkage and creep may be very significant.

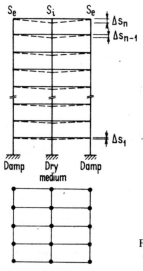

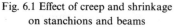

Fig. 6.1 Effect of creep and shrinkage
on stanchions and beams

If all the columns of a frame structure are placed in the same ambient conditions, and if they are equally loaded and have the same degree of reinforcement, then no changes of the internal forces occur due to shrinkage and creep. Actually, the cross-sections of the internal columns are usually larger, their degree of reinforcement is higher and their ambient conditions differ from those of the external columns. Consequently, the columns shorten differently, because the greater loaded columns have larger creep deformations, the more reinforced columns deform less than those with less reinforcement, and the effects of creep and shrinkage are more pronounced in a dry environment (i.e. for internal columns). Hence, the internal forces change. An example is illustrated in Fig. 6.1, where the shortening of the internal column S_i, relative to the peripheral columns S_e, is Δs_n; this shortening reaches its highest value in the top floor. The resulting deformation of the horizontal members is indicated by dashed lines.

A very similar situation can occur in tall structures with columns and a reinforced concrete shear wall or lift shaft (Fig. 6.2). If this shaft is cast earlier than the columns, the columns deform more due to shrinkage and creep than the bracing member, and the internal forces undergo a change; the deformation of the horizontal

265

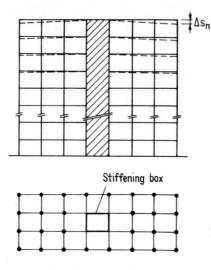
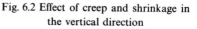

Δs_n

Stiffening box

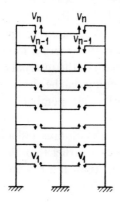

Fig. 6.2 Effect of creep and shrinkage in
the vertical direction

members again is shown by dashed lines. If, however, the columns are made of steel and the bracing member of reinforced concrete, only this latter part of the structure is affected by shrinkage and creep, and the horizontal members deform in the opposite sense.

The shortening of the columns has to be determined for each floor separately, because the effects of creep and shrinkage vary in the course of the construction process.

The magnitude of the shortening Δs_n may be demonstrated by an example in which a structure with 45 floors ($n = 45$) is considered; the strain difference between the internal and external columns is 150×10^{-6}. With the height of one floor equal to $3\,200$ mm, the shortening of the internal columns relative to those on the periphery of the structure is $\Delta s_n = 150 \times 10^{-6} \times 3\,200 \times 45 = 21.6$ mm at the n-th level. Evidently, such a displacement changes the internal forces considerably — the more so, the taller the structure. For example, the shear forces V in the beams of a symmetrical structure are shown in Fig. 6.3.

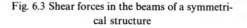

Fig. 6.3 Shear forces in the beams of a symmetrical structure

In low structures, shortening of the columns due to shrinkage and creep may be usually neglected. However, it is recommended, for structures with less than 30 floors and with not too rigid horizontal members (for example, with larger spans, or in the case of lift-slab floors), to determine the shrinkage and creep deformations approximately (e.g., by the method employed in the above example). Such practice is useful, because the calculated values give a realistic idea for the design of the structural details, such as partition walls or linings. These items are extremely sensitive to deformations (Fig. 6.4), because the partition wall is deformed due to the different compression of the columns — it becomes separated from the adjoining structural members and cracks develop in its corners.

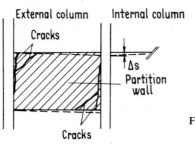

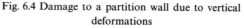

Fig. 6.4 Damage to a partition wall due to vertical deformations

Attention should be paid to the non-loaded members of a structure, viz. the linings and the panels on the faces of buildings. Owing to a shortening of the columns caused by shrinkage and creep, these members buckle until they are completely separated from the load-bearing structure; the damage is increased in summer season when the ambient temperature is high. These consequences can be avoided if the lining and facing members are allowed free motion, i.e. no stresses from the load-bearing structure are transferred to them.

When a structure has more than 30 floors, or when its horizontal members are rigid, the effect of the different shortening of the vertical members must be ascertained more accurately.

Deflections of beams

The influence of creep in the calculation of the deflections of reinforced concrete structures with and without cracks was discussed in the previous Chapters, but shrinkage also may influence the magnitude of the deflection of beams with asymmetrical reinforcement.

It has been shown that the strain caused by shrinkage is reduced in the regions with a high degree of reinforcement as compared with regions having less reinforcement. If the member is reinforced symmetrically, this strain is uniform over the entire depth of the section (actually, the strain is larger in the middle part of the section with no reinforcement). However, if the member is reinforced only

267

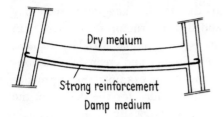

Fig. 6.5 Deformation of a simply reinforced beam due to shrinkage

in its tensioned part, it deflects due to shrinkage, and, for members having a larger depth, this deflection can be increased by a difference in the ambient conditions (Fig. 6.5) (the upper surface of a beam may be exposed to drier conditions than its lower surface). Generally, partition walls supported on beams can be damaged due to the deflection of these beams and the effect is augmented by creep and shrinkage. Such damage is illustrated in Fig. 6.6: cracks in the partition develop along the columns and in the corners. Cracks can be eliminated or reduced by providing a layer of elastic material between the bottom of the wall and the top of the beam. Also, the partition wall should be built as late as possible after the casting of the beam so that in the intervening period, some of the shrinkage and creep can develop beforehand.

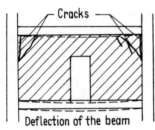

Fig. 6.6 Damage to a partition wall due to the deflection caused by creep

Effects in the horizontal direction

In the horizontal direction of floor structures, the effect of shrinkage prevails, because generally they are not subjected to any horizontal stresses (unless they are prestressed). However, stresses will develop due to shrinkage if the horizontal structures are fixed in the shear walls or shafts. Such an arrangement prevents the free development of the strain caused by shrinkage, and tension appears in the horizontal members. Such a case is pictured in Fig. 6.7a, where a tensile force N acts in the beams; this force must be considered in the analysis of the structure. To diminish this effect of shrinkage, it is sometimes recommended to leave out strips 600 to 900 mm wide in the concrete floor slab and to fill them with concrete mix a few weeks later, when a certain part of the strain caused by shrinkage had already developed. In these strips, the reinforcement bars are separated and

a)

Stiffening walls

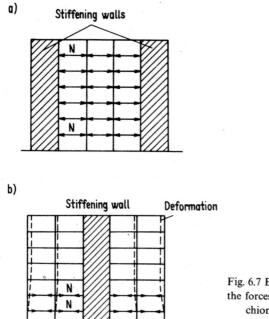

b)

Stiffening wall   Deformation

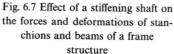

Fig. 6.7 Effect of a stiffening shaft on the forces and deformations of stanchions and beams of a frame structure

reconnected prior to the additional casting. However, the magnitude of the tensile force and the appropriate reinforcement must be calculated in any case.

If the bracing element (wall or shaft) is situated in the middle of the structure (Fig. 6.7b), the forces in the horizontal members are small or zero; the shortening of the beams deforms the columns in which bending moments develop. Moreover, in the lowest floors, tensile forces are produced as a result of the fixed columns in the foundations.

6.1.2   Panel structures

Structures made of panels are erected when a large part of shrinkage had already developed. In spite of this, there appear structural and visual defects in these structures that must be accountable to the long-term volume changes, i.e. to shrinkage and creep; also, they may be augmented by the influence of temperature. It is characteristic for prefabricated structures that members of different ages may be mounted side by side and connected. According to observations made on sites, the average difference in the age of concrete of neighbouring panels is 14 days, the maximum being 54 days. Obviously, cracks are bound to develop, especially in the connections.

Effects of shrinkage

If the strain due to shrinkage cannot develop freely, large tensile stresses develop in the panels and in the connections between them so that cracks may appear. They form especially in the vertical joints, in the connections between the wall and floor panels and in the connections of the vertical panels to the foundations.

In the direction *perpendicular to the load-bearing panel walls,* the floor slabs tend to shrink and the wall panels are displaced perpendicularly to their plane. A similar situation is shown in Fig. 6.7b, where the displaced panels are drawn by dashed line. Bending moments develop in the floor slab — bearing wall connection, because the floor panels never rest completely freely on the wall panels. Other changes in the state of stress in the wall panels are caused by the change of their original vertical position. A very adverse situation arises when the bracing elements are situated at the extremities of the structure (refer to Fig. 6.7a), because the tensile forces in the floor panels may cause a failure of the connection with the wall panels.

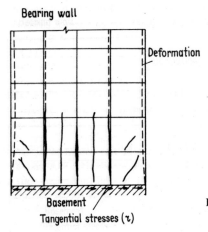

Bearing wall

Deformation

Basement

Tangential stresses (τ)

Fig. 6.8 Effect of shrinkage of the foundations on a panel wall

No stresses can develop in the plane of the load-bearing walls if strains due to shrinkage can develop freely in all floors. Such a condition, however, cannot exist in the ground floor, where the wall panels are attached to the foundations. Shear stresses form in this connection which affects the original state of stress in the adjacent floors. Then cracks appear in the vertical connections between the panels and even in the panels themselves if they are not strong enough to resist the tensile stresses (Fig. 6.8). The shear stresses in the panel-foundation connections increase towards the end of the panel wall and so does the probability of the formation of cracks. The shear stresses, together with the normal stresses acting vertically, produce cracks perpendicular to the direction of the principal tensile stresses.

270

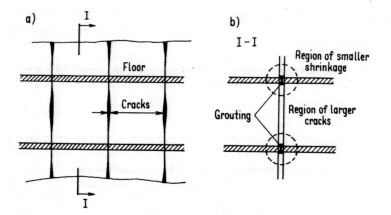

Fig. 6.9 Effect of shrinkage on wall panels

Vertical cracks between the wall panels are not excluded even in higher floors, if these panels shrink differently in the individual floors at the top, at the bottom and in the middle of the floor height. The reasons for this difference in shrinkage are:

(a) The fictitious perimeter, on which the intensity of shrinkage depends, is less at the connection of the wall with the floor panels than in the rest of the panel, and hence shrinkage develops with less intensity at that position (Fig. 6.9b); shrinkage of the panel at these joints is usually considered to be 10 to 30 per cent smaller than in the remaining panel;

(b) the mortar and grout between the heads of the panels transmit moisture to the adjacent heads of the connected panels for a long period of time;

(c) the air flow in the corners of the rooms (i.e. at the joints of the floor and wall panels) is less than around the surface of the middle part of the wall panels, hence, the drying of these parts is more intensive;

(d) the grouted joints are almost always provided with longitudinal reinforcement which reduces shrinkage.

Cracks caused by the above effects develop in the joints between the panels and are largest in the middle of the floor height. Their width may reach several tenths of millimetres and they are visible after about one year following the completion of the structure (Fig. 6.9a).

The load-bearing panel walls resist the vertical load consisting of their dead weight and that of the floor structure and of the live load, but they also have to take the horizontal loads (such as wind, seismic forces, etc.). For these loads, the panel wall is considered as a monolithic structure, and the joints between the panels must be able to transfer all the stresses. If these joints are cracked, the assumed static behaviour of the wall is changed so that the wall acts as one composed of vertical panel strips and its resisting capacity is reduced. For this reason, the effects of shrinkage have to be taken into account and an appropriate structural arrangement

271

must ensure the behaviour of the wall as a whole; this purpose is attained, for example, when the floor and wall panels are bonded so that the joints between them are discontinuous.

Shrinkage of concrete (usually acting together with creep) may induce vertical stresses in the panels of the load-bearing walls; if the wall panels do not act independently and are of different ages so that they shrink differently, shear stresses develop in the vertical joints and the stresses due to the vertical load are transferred from the younger panels to the older ones. Vertical cracks between the panels can also form if the grouting has an insufficient load-bearing capacity and if the panels are not designed to resist vertical displacement (notches, etc.). If the tops of the wall-panels are stiffened by a cast-in-situ lintel, additional stresses due to the above effects also develop in this lintel.

Floor panels are similarly affected by shrinkage when the neighbouring panels are of different age. Tensile stress develops in the panel with the larger shrinkage and the stresses in the supports on the walls increase. The normal forces N and the shear forces Q originated by shrinkage are depicted in Fig. 6.10.

With regard to the visual effect, cracks *between the adjacent floor panels* merit attention; they are caused by a difference in shrinkage of the individual panels

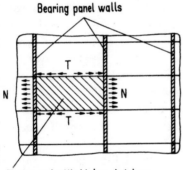

Fig. 6.10 Effect of shrinkage on floor panels

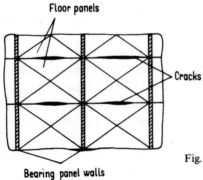

Fig. 6.11 Cracks between the neighbouring floor panels caused by shrinkage

resulting in tensile stresses in the joints and, eventually, cracks. These cracks are nearly always present and they are widest at the mid-span of the panels (Fig. 6.11). They may also reflect on the static behaviour of the floor structure, because, due to the cracked connection, the panels cannot co-operate in resisting the vertical load. Hence, the joints should be designed carefully and the grout should be of high quality.

Effects of creep

Creep of concrete may have an adverse effect in the following cases:

(a) Parts of a statically indeterminate structure are subjected to different strains due to creep.

(b) The structural system changes in the course of the construction process.

(a) A panel floor or a panel wall is considered as monolithic for the purpose of analysis and is calculated as a statically indeterminate structure, but the individual panels creep differently on account of differences in material or age of concrete. If two equally loaded adjoining panels have different creep properties, their deflections will also differ, assuming that they can deflect freely. If this is not possible, shear forces develop in the joint and the older panel takes a higher load. As the strength of the grouting may be reduced due to shrinkage, the joint may fail and the two panels cannot co-operate.

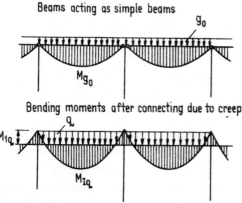

Fig. 6.12 Change of structural system

(b) The change of the structural system is not as significant in panel structures as it is in industrial buildings and bridges, but such a change occurs to some extent even in the panel structures. During erection, the floor panels are simply supported by the load-bearing panel walls (Fig. 6.12), and they behave as statically determinate members loaded by their dead weight. After the completion of construction, however, they are partly clamped in the wall panels of the lower and upper floors so that the structural system changes and bending moments due to

273

creep start to develop at the support sections. These sections may then fail, especially if tensile stresses due to a difference in shrinkage have developed simultaneously in the floor panels (Fig. 6.10).

6.2   Industrial buildings

With regard to the effects of creep and shrinkage, these buildings are divided into those with cast-in-situ concrete and those erected from precast concrete members.

6.2.1   Cast-in-situ structures

The effects of creep and shrinkage on the state of stress in these structures are very similar to those discussed in Section 6.1.1, if the structure is designed as a concrete frame. Shrinkage has to be considered especially in the horizontal direction, while it can be usually neglected in the vertical direction. The beams of the frame are shortened due to shrinkage by Δs (Fig. 6.13) and the original state of stress changes (tension develops in the beams and additional bending moments in the stanchions increase with time); the internal forces may be augmented by the influence of temperature. Also, creep and shrinkage increase the deflection of the beams (see Fig. 6.5).

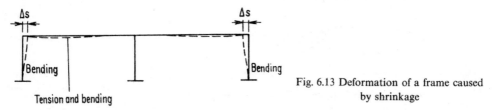

Fig. 6.13 Deformation of a frame caused by shrinkage

Watertanks, silos and bunkers need to be analysed as three-dimensional structures. The bottom of such structures is often cast much earlier than the walls, and consequently, the shrinkage strains of the bottom ε_{s1} and of the walls ε_{s2} may be very different. Since the walls cannot move freely (Fig. 6.14), additional bending

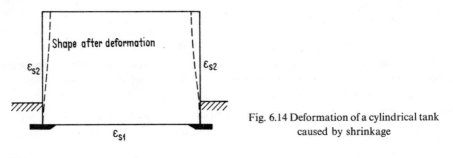

Fig. 6.14 Deformation of a cylindrical tank caused by shrinkage

moments and tensile forces develop at the heel of the walls and along the periphery of the bottom. This may endanger the normal serviceability of the structure, especially if liquids are stored.

Creep of concrete shells (especially roofs) may increase the deflection of such structure to such an extent that it may buckle.

To express the effect of creep on the stability of shells with the necessary safety margin, it is recommended [6.1] to use a reduced modulus of deformability in the analysis, i.e.

$$E_{cd} = \frac{E_c}{3}$$

where E_c is the modulus of elasticity. Furthermore, this modulus is reduced by 25 per cent to allow for the heterogeneity of concrete, hence

$$E'_{cd} = \frac{E_c}{4}$$

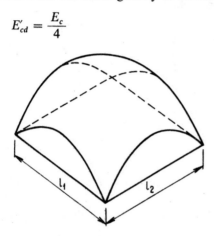

Fig. 6.15 Shell of double curvature

The recommendations apply to approximate analyses of the stability of shells together with the following relations:

For circular shells and for those of double curvature (Fig. 6.15), the full load $q\ [\mathrm{kNm^{-2}}]$ should not exceed the value

$$q_0 = 0.2E'_{cd} \left(\frac{d}{R}\right)^2$$

where d is the thickness of the shell;

R is the radius of curvature of a circular shell measured at the top of the shell.

For shells of double curvature

$$R \doteq \sqrt{R_1 R_2}$$

where R_1 and R_2 are the radii of the main curvatures. The total load P (for example in kN) should not exceed the value

$$P_0 = 0.4\pi E_{cd} d^2$$

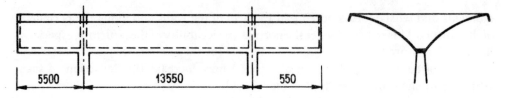

Fig. 6.16 Long cylindrical shell

For long cylindrical shells without stiffeners (Fig. 6.16), the normal stress σ_x should not exceed the value

$$\sigma_0 = 0.25E'_{cd}\frac{d}{R}$$

where d is the thickness of the shell;
$\quad\quad\quad R$ is the radius of curvature;
$\quad\quad\quad \sigma_x$ is the normal stress along the X-axis.

The shear stress τ in these shells should be less than

$$\tau_0 = 0.3E'_{cd}\left(\frac{d}{R}\right)^{\frac{2}{3}}$$

and both stresses must satisfy the condition

$$\frac{\sigma}{\sigma_0} + \left(\frac{\tau}{\tau_0}\right)^3 \leqq 1$$

where σ and τ are normal and shear stresses calculated on the basis of the theory of elasticity.

If the above relations are not satisfied, stiffeners have to be used with maximum spacing equal to $7\sqrt{Rd}$.

For short cylindrical shells without stiffeners (Fig. 6.17), the load $q\,[\text{kNm}^{-2}]$ should not exceed the value

$$q = 0.75E'_{cd}\left(\frac{d}{R}\right)^2\frac{1}{\dfrac{l_1}{\sqrt{dR}} - 1}\quad\quad[\text{kN m}^{-2}]$$

where l_1 is the distance of the supporting arches.

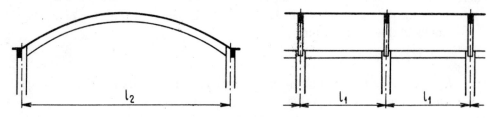

Fig. 6.17 Short cylindrical shell

6.2.2   Structures assembled of precast members

Creep and shrinkage have to be considered in these structures even though some of the volume changes have developed prior to the mounting of the precast members in the structure. Precast structures have their special problems, and even small effects of creep and shrinkage may have serious consequences.

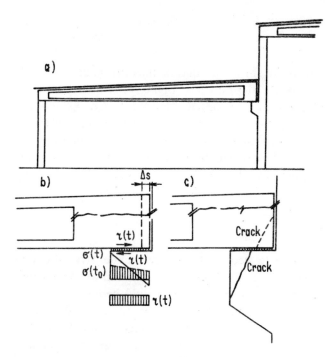

Fig. 6.18 Stresses and damage caused by creep and shrinkage

Thus, the change of the conditions at the heads of precast beams and around their supports due to shrinkage and creep may cause serious damage to the structure. This is demonstrated on a precast roof beam (Fig. 6.18a) resting in a mortar bed on reinforced-concrete cantilevers. The beam is placed into position at time t_0 and subjected to service load. The stress diagram for $\sigma(t_0)$ at the beam-cantilever joint is shown in Fig. 6.18b, the shear stress $\tau(t_0) = 0$. In the period between t_0 and t, the deflection of the beam increases due to creep and a redistribution of stress $\sigma(t)$ follows with the new resultant $R(t_0)$ moving very close to the internal edge of the supporting cantilever. This concentration of stress may cause cracks in the support (indicated by the solid line). Moreover, shrinkage shortens the beam and induces shear stresses $\tau(t)$ at the joint of the beam with the support. These stresses may also be the cause of cracks in the beam (dashed line of Fig. 6.18c). Both these kinds of cracks may be so serious that they need attention, and such repairs are usually costly and difficult to execute.

277

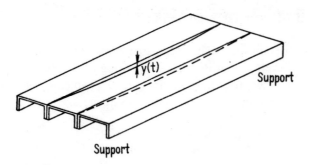

Fig. 6.19 Differences in floor panels
caused by creep

Creep (with some contribution due to shrinkage) may be the cause of unfavourable differences in the deflections $y(t)$ of the floor panels (Fig. 6.19). This occurs most frequently when panels of different age are mounted side by side. Cracks, induced by shear stresses developed in the panel joints cause the panels to take the load separately. Also, the flooring and the floor insulation may suffer damage.

The redistribution of the internal forces, caused by the eventual changes of the structural system, must also be considered (Fig. 6.12).

6.3   Bridges

In modern bridges which are characterized by large spans, slender structures and special construction methods, the volume changes of the material are taken into account by the usual analysis. However, this approach should be also applied to smaller bridges, whose cross-sectional dimensions tend to be less and stresses higher. Neglecting of the long-term effects of shrinkage and creep may be the reason for subsequent excessive deflections which devaluate the structure visually and reduce its serviceability. Any attempts to re-establish the original state of the bridge are very costly, if not impossible. Besides the deflection of a bridge, which is very important for traffic, a redistribution of stresses may occur, which has a maximum effect in continuous structures cast in situ. A pre-knowledge of the development of deflection is essential during the construction of a bridge of precast members by segmental cantilevering, which again depends on appropriate knowledge of creep and shrinkage. With piers of very different heights and cast at different times, their shortening due to shrinkage and creep may cause a change of stresses in the upper horizontal structure, which must be considered quantitatively in the analysis.

The volume changes may be significant in bridges erected of precast and prestressed girders mounted side by side. If they are of different age, the stress transfers from the younger girders to the older ones. Such a situation may arise when the bridge has to be widened by adding new girders to the old structure.

In strongly reinforced girders, sometimes cracks develop perpendicularly to the longitudinal axis of the member due to shrinkage (Fig. 6.20). Such cracks are wider

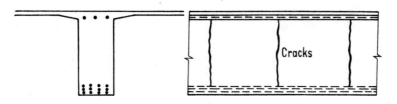

Fig. 6.20 Long-term deflection of frame bridges

at the mid-depth but narrower at the top and bottom surface of the member. This is explained by the heavier reinforcement at the upper and lower parts of the cross-section and by less or no reinforcement in the middle region, because the reinforcement restrains shrinkage. Cracks of this kind are not dangerous for the load-bearing capacity of the girder as long as they do not develop at sections with high shear stresses or torsional moments; however, they allow water to penetrate to the steel bars and thus facilitate their corrosion. Hence, such sections should also be reinforced longitudinally in the middle region.

The governing criterion, however, is the magnitude of the deflection of a bridge. Long-term observations of bridges of various structural systems have been carried out for this reason. The results of these observations furnished the possibility of determining the absolute values of deflections and of evaluating the quality of a bridge structure. Let us quote examples of two prestressed bridges (Fig. 6.21), where the deflection was measured at the middle of the largest intermediate span. Both bridges have the same structural system, their spans do not differ very much and the same erection method was used in both cases. The structure of both bridges consists of a continuous girder with three spans and with a hinge in the intermediate span. The bridge depicted in Fig. 6.21a has spans of 42.50, 70.00 and 42.50 m and the observations lasted 15 years, the other bridge (Fig. 6.21b) has spans of 35.10, 61.60 and 31.50 m and was observed for 8 years. The comparison of the time-dependent diagram of deflection indicates that the difference in the behaviour of both bridges is quantitative and qualitative. The larger bridge deflects much more than the smaller one: after eight years, the deflection of the larger bridge was 300 mm, while the deflection of the smaller bridge was only 70 mm after the same period of time; the deflection of the first bridge is still increasing, while that of the second bridge has stabilized. The deflection of the first bridge has already impaired its serviceability (deformed railing).

The deflections of other two bridges are shown in Fig. 6.22; both are frame structures of the same span but the quality of the concrete of the first bridge is inferior. The magnitude of the deflection of the first structure was 140 mm after 20 years of observation, while the deflection of the other bridge was only about 50 mm. Also, in this case, the growth of the deflection of the second bridge practically ceased after five years, while the deflection of the first bridge continued to increase. Even if the total deflection of the two bridges cannot be ascribed solely

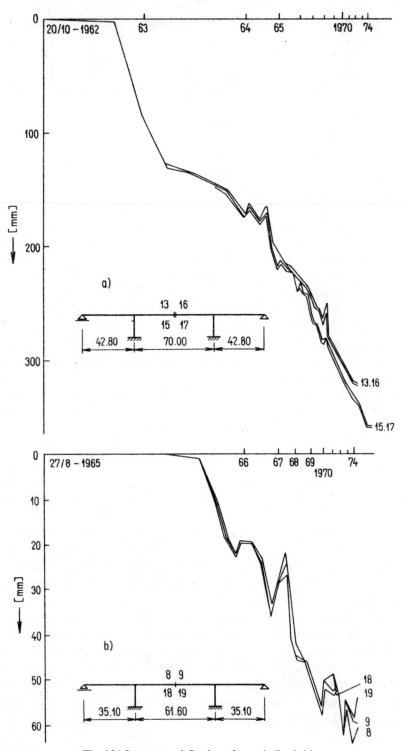

Fig. 6.21 Long-term deflection of two similar bridges

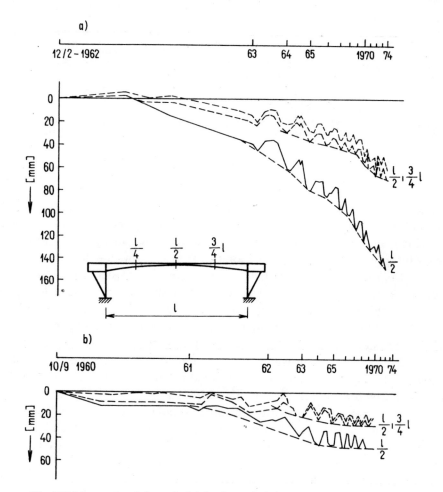

Fig. 6.22 Measured and theoretical deflections of a bridge assembled from segments

to creep, because part of it is due to the change of the horizontal force, clearly, the effect of creep is predominant.

In all the four bridges, the measured deflection under long-term loading cannot be compared with the theoretical deflection, because, at the time of their design, no accurate analyses of the effect of creep were undertaken, and the data needed for a new analysis (the development of loading during the construction, prestressing, etc.) are not available any more; without them, it is impossible to determine the deflection induced by creep and shrinkage accurately.

For some recently constructed bridges, the long-term effects have been analysed, using the rate-of-creep theory for the creep prediction, and the measurements of deflections are continuing. Let us compare the theoretical and the measured deflections of a prestressed bridge designed as a continuous girder with three spans.

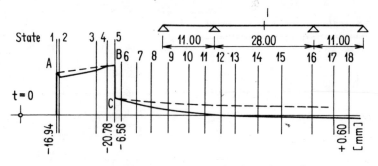

Fig. 6.23 Theoretical and measured values of φ and ε_s

The bridge has been assembled from segments arranged on scaffolding and post-tensioned by prestressing cables. The segments, having the width of the bridge and a length of 3 m (in the direction of the longitudinal bridge axis), were glued together by epoxy resin in the contact joints. Fig. 6.23 shows the deflection diagram at the mid-point of the intermediate span; the measured deflections are shown by solid line, while the theoretical deflections by dashed line. The deformations (a camber in this case) in the stages 1 to 5 of the construction were caused by the provisional and final prestressing, when the segments were loaded only by their dead weight. From stage 5, the bridge was subjected to the full load. It may be seen from the figure that the measured deflection induced by creep (and shrinkage) increases more rapidly in time than the theoretical deflection: for example, the deflection measured at the time of the 14th stage (200 days) was $+0.11$ mm, whereas the theoretical deformation (camber) should have been -4.00 mm. This difference is due mainly to the inaccuracy of estimating those creep and shrinkage factors which are dependent on the ambient conditions; these factors are important especially in the first stages of the construction (approx. stages 5 to 11).

The uncertainty of these estimates is illustrated in Fig. 6.24 which shows the theoretical time distribution of the factors φ and ε_s in accordance with the CEB-FIP Recommendations: dashed lines designate the recommended values and solid lines those calculated from the measurements. The measurements registered in Figs 6.24b and 6.24c were obtained in observations carried out in Czechoslovakia.

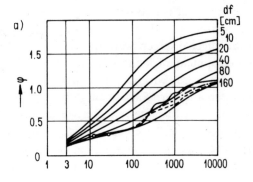

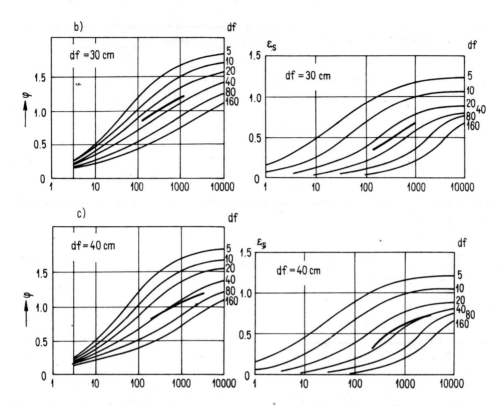

Fig. 6.24 Cracks caused by shrinkage of a strongly reinforced beam

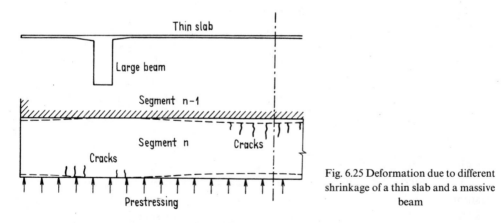

Fig. 6.25 Deformation due to different shrinkage of a thin slab and a massive beam

Shrinkage can be the cause of cracks during the prestressing of the segments, when the epoxy glue is in liquid state and acts as a lubricant. This may be explained by the difference in shrinkage of the individual parts of the segment, because the relatively thin slab dries more rapidly than the massive beam and the segment is deformed (see Fig. 6.25). The joint, at which two adjoining segments

283

should contact over their entire faces, is deformed by the projecting (less shrinking) beams establishing mutual contact; when the segments are mounted and prestressed, these projecting heads lean against each other (Fig. 6.26), the segments deform and cracks develop. The value of Δs can be as much as 0.40 mm, in addition to which the deformations are added as the assembly and prestressing progress.

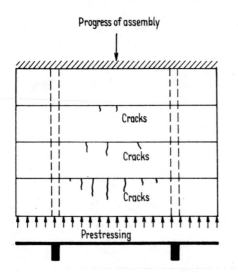

Fig. 6.26 Cracks due to prestressing of each segment

It is evident from the foregoing discussion that an accurate prediction of creep and shrinkage is very difficult; hence, it is advisable to determine their lower and upper limits and to use these extreme values in the evaluation of their effects on the static behaviour of the bridge structure.

To analyse the effects of creep on the changes of the internal forces in the direction of the longitudinal bridge axis, any of the previously discussed methods may be employed by considering the effects of dead load, prestressing forces and other loadings, and the forced deformations. However, an approximate method for the evaluation of the effect of creep on the bending moments may be applied to the analysis, at least for checking the structure. The method based on the application of the rate-of-creep theory consists of the following steps:

(1) The bending moments induced by the above-mentioned loads are calculated for all the structural systems involved during the construction process;

(2) The bending moments induced by the above loads are determined under the assumption of the structure being constructed all in one operation, i.e. by ignoring the different structural systems existing during the construction process;

(3) The difference beween the bending moments ascertained in steps (1) and (2) is determined;

(4) The bending moments obtained in step (3) (their diagram being always linear) are multiplied by $(1 - e^{\Delta\varphi})$, where $\Delta\varphi$ is the creep coefficient corresponding to the investigated period; the resulting moments are those caused by creep;

284

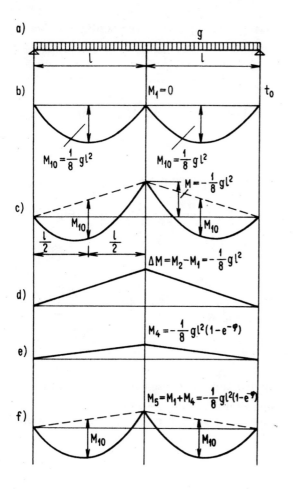

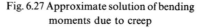

Fig. 6.27 Approximate solution of bending moments due to creep

(5) The sum of the bending moments obtained in steps (1) and (4) yields an estimate of the resulting moments including the effect of creep.

Let us determine, as an example, the support bending moments of a continuous beam with two equal spans l (Fig. 6.27a). At time t_0, the structure consists of two simple beams loaded by their dead weight g; at time t, the beams are connected by concreting the gap over the intermediate support, at which point a redistribution of the bending moments occurs.

(1) The bending moment at the mid-spans of the two simple beams (Fig. 6.27b) is

$$M_{10} = \frac{1}{8} gl^2$$

The bending moment over the intermediate support is

$$M_1 = 0$$

(2) The bending moment at the intermediate support of the continuous beam (Fig. 6.27c) is

$$M_2 = -\frac{1}{8}gl^2$$

(3) The difference between the support bending moments M_1 and M_2 is

$$M_2 - M_1 = -\frac{1}{8}gl^2$$

The diagrams of the difference of the bending moments in the spans are linear (Fig. 6.27d).

(4) The diagrams of the bending moments due to creep are also linear (Fig. 6.27e). The support bending moment caused by creep is

$$M_4 = -\frac{1}{8}gl^2(1 - e^{-\varphi})$$

The creep coefficient φ is determined with regard to the times t and t_0.

(5) The total support bending moment (Fig. 6.27f) is

$$M_5 = M_1 + M_4 = -\frac{1}{8}gl^2(1 - e^{-\varphi})$$

The value of the bending moment at the support caused by the effect of creep lies between that of the continuous and simply supported beams. It means that the role of creep, originating after the change of the structural system, is to adapt the forces to this new structural system.

References

Chapter 1 — References

[1.1]   Recommandations Internationales CEB–FIP pour le Calcul et l'Exécution des Ouvrages en Béton. Bulletin d'Information, 1970, No. 72, Paris.

[1.2]   Hansen, T. C.: Creep of Concrete. Swedish Cement and Concrete Research Institute at the Royal Institute of Technology. Bulletin 33, 1958, Stockholm.

[1.3]   Illston, J. M.: The Components of Strain in Concrete under Sustained Compressive Stress. Magazine of Concrete Research 17, 1965, No. 50.

[1.4]   Rüsch, H. et al.: Valutazione critica dei procedimenti per stabilire l'influenza della viscosità è del ritiro del calcestruzzo sul comportamento delle strutture portanti. Costruzioni in cemento armato. Italcement, 1974, No. 11, Milan.

[1.5]   Neville, A. M.: Theories of Creep in Concrete. ACI Journal, Proc. 52, 1955.

[1.6]   Gvozdyev, A. A.: Creep of Concrete in: Mechanics of solid body. (In Russian.) Stroyizdat, Moscow, 1966.

[1.7]   Gvozdyev, A. A.: Le Fluage du Béton. 2me Congrès de Méchanique Théorique et Appliquée. CEB, 1964, Moscow.

[1.8]   Wagner, R. D.: Das Kriechen unbewehrten Betons. Deutscher Ausschuss für Stahlbeton 1958, No. 131, Berlin.

[1.9]   Skudra, A. M.: Résistance à longterme du béton à la traction. Academy of Sciences of Latvian SSR, Riga, 1956.

[1.10]   Korsak, N. D.: Investigation of strength and elastic properties of concrete. Strength, elasticity and creep of concrete. (In Russian.) Gosstroyizdat, Moscow, 1941.

[1.11]   Bushkov, V. A.: Reinforced concrete structures. (In Russian.) Stroyizdat, Moscow, 1940.

[1.12]   L'Hermite, R.: Technological problems of concrete. (In Russian.) Gosstroyizdat, Moscow, 1959.

[1.13]   Prokopovich, J. E.: Effect of long-term processes on stressed and deformed state of structures. (In Russian.) Gosstroyizdat, Moscow, 1963.

[1.14]   Meyer, H. G.: On the Influence of Water Content and of Drying Conditions on Lateral Creep of Plain Concrete. Materials and Structures 2, 1969, No. 8.

[1.15]   Gopalakrishnan, K. S. et al.: Creep Poisson's Ratio of Concrete under Multiaxial Compression. ACI Journal, Proc. 66, 1969, No. 12.

[1.16]   Backström, S.: Creep and Creep Recovery of Cement Mortar. IABSE, 5th Congress, Preliminary Report, 1956, Lisbon.

[1.17]   Klokner, F.: Elasticity tests of concrete. (In Czech.) Zprávy veřejné služby, 1933, Praha.

[1.18]   Ulickiy, I. I. et al.: Stiffness of members subjected to bending. (In Russian.) Akademia stroitelstva i arkhitektury USSR, Kiev, 1963.

287

[1.19]   Alexandrovskiy, S. V.: Analysis of concrete and reinforced concrete structures for changes of temperature and humidity with consideration of creep. (In Russian.) Stroyizdat, Moscow, 1973.

[1.20]   Ross, A. D.: Creep of Concrete under Variable Stress. ACI Journal, Proc. 54, Vol. 29, 1958, No. 9.

[1.21]   Die materialtechnischen Grundlagen und Probleme des Eisenbetons in Hinblick auf die zukünftige Gestaltung der Stahlbeton-Bauweise. EMPA, Bericht 162, 1952, Zürich.

[1.22]   Hummel, A. et al.: Einfluss der Zementart, des Wasser–Zement-Verhältnisses und Belastungsalters auf das Kriechen des Betons. Deutscher Ausschuss für Stahlbeton, 1962, No. 146.

[1.23]   Ulickiy, I. I.: Theory and analysis of reinforced concrete frame structures with consideration of long-term processes. (In Russian.) Budivelnik, Kiev, 1967.

[1.24]   L'Hermite, R. G. – Mamillan, M.: Further Results of Shrinkage and Creep Tests. Intern. Conf. on the Structures of Concrete. Cement and Concrete Association, London, 1968.

[1.25]   US Bureau of Reclamation: A Ten-Year Study of Creep Properties of Concrete, Concrete Laboratory Report No. SP–38, 1953.

[1.26]   Hummel, A.: Vom Einfluss der Zementart, des Wasserzementverhältnisses und des Belastungsalters auf das Kriechen von Beton. Zement–Kalk–Gips 12, 1959, No. 5.

[1.27]   Rüsch, H. et al.: Der Einfluss des mineralogischen Charakters der Zuschläge auf das Kriechen von Beton. Deutscher Ausschuss für Stahlbeton, 1962, No. 146.

[1.28]   Kordina, K.: Experiments on the Influence of the Mineralogical Character of Aggregates on the Creep of Concrete. RILEM, Bul. No. 6, 1960.

[1.29]   Troxell, G. E. et al.: Long-time Creep and Shrinkage Tests of Plain and Reinforced Concrete. ASTM Proc. 58, 1958.

[1.30]   Guide to Structural Lightweight Aggregate Concrete. ACI Journal 1967, 8.

[1.31]   Kruml, F.: Long-term deformation properties of lightweight concretes. (In Slovak.) Stavebnícky časopis 13, 1965, No. 3, Bratislava.

[1.32]   Structures en béton léger. Bulletin d'Information CEB, 1972.

[1.33]   Development of lightweight prestressed bridges of lightweight concretes. (In Czech.) VÚIS Bratislava, Report on the problem No. P 12–526–072/3.4, 1979.

[1.34]   Wallo, E. M. et al.: Sixth Progress Report: Prediction of Creep in Structural Concrete from Short-time Tests. Univ. of Illinois. T. and AM Report No. 658. 1965.

[1.35]   Hannant, D. J.: Strain Behaviour of Concrete up to 95 °C under Compressive Stresses. Conf. on Prestressed Concrete Pressure Vessels. Inst. of Civil Eng., Paper 17, 1967.

[1.36]   ACI Committee 517. Low Pressure Steam Curing. ACI Journal, Proc. 60, 1963.

[1.37]   Hanson, J. A.: Prestress Loss as Affected by Type of Curing. Prestressed Concr. Inst. Journal, 9, 1964.

[1.38]   L'Hermite, R.: Nouveaux résultats de recherches sur la déformation et la rupture du béton. Ann. Inst. Techniques du bâtiment et des travaux publics, 28, No. 207 – 208, 1965, Paris.

[1.39]   Le Camus, B.: Recherches experimentales sur la déformation du béton et du béton armé. Déformations lentes. Paris, Inst. Technique du bâtiment et des travaux publics, 1947.

[1.40]   Andersen, P.: Experiments with Concrete in Torsion. Am. Soc. of Civil Engs, Transactions, 100, 1935.

[1.41]   Probst, E.: Plastic Flow in Plain and Reinforced Concrete Arches. ACI Journal, Proc. 30, 1933.

[1.42]   Probst, E.: The Influence of Rapidly Alternating Loading on Concrete and Reinforced Concrete. The Structural Engineer, 9, 1931.

[1.43]   Creep of bridges (In Czech.) VÚIS Bratislava, Report on problem No. P 12–526–072/2.5, 1975.

[1.44]   Šmerda, Z.: Theoretical and practical problems of construction materials and their application to structures – 2nd part. (In Czech.) Dom techniky SVTS, Bratislava, 1974.

[1.45]   Le Camus, B.: Recherches sur le comportement du béton et du béton armé soumis à des efforts répétés. Inst. Technique du bâtiments et des travaux publics. Circulaire Série F, No. 27, 1946.

[1.46]   Gvozdyev, A. A. et al.: On the deformations of concrete under many times repeated load. Behaviour of static and dynamic structures. (In Russian.) Stroyizdat, Moscow, 1972.

[1.47]   Kulygin, J.: Creep of concrete under many times repeated compressive loads. Peculiarities of concrete deformations. (In Russian.) Moscow, 1969.

[1.48]   Kulygin, J. – Belobobrov, I. K.: Experimental investigation of creep of concrete under many times repeated cyclic loading. Strength and rigidity of reinforced concrete structures. (In Russian.) Moscow, 1968.

[1.49]   American Concrete Institute. Building Code Requirements for Reinforced Concrete. ACI Com. 318–63, 1963, Detroit.

[1.50]   Arutunyan, N. Kh.: Some problems of creep theory. (In Russian.) Gostechteoretizdat, Moscow, 1952.

[1.51]   Hummel A.: Das Beton ABC, Berlin, 1959.

[1.52]   Rüsch, H. – Jungwirth, D.: Stahlbeton. Berücksichtigung der Einflüsse von Kriechen und Schwinden auf das Verhalten der Tragwerke. Werner–Verlag, Düsseldorf, 1976.

Chapter 2 – References

[2.1]   Bažant, Z. P.: Theory of Creep and Shrinkage in Concrete Structures – A Précis of Recent Developments. Mechanics Today, Vol. 2, Pergamon Press, 1975.

[2.2]   Bažant, Z. P. – Wittmann, F. H.: Creep and Shrinkage in Concrete Structures. J. Wiley and Sons, Chichester–New York–Brisbane–Toronto–Singapur, 1982.

[2.3]   Bažant, Z. P. – Wu, T. S.: Dirichlet Series Creep Function for Aging Concrete. Journal of the Engineering Mechanics Div., ASCE, Vol. 99, Proc. Paper 9645, April 1973, pp. 367 – 388.

[2.4]   Rüsch, H. et al.: Valutazione critica dei procedimenti per stabilire l'influenza della viscosità e del ritiro del calcestruzzo sul comportamente delle strutture portanti. Costruzioni in cemento armato, No. 11, Italcementi 1974, Milan.

[2.5]   Štrunc, A.: Some Problems of Creep Analysed by Time Discretization Method. (In Czech.) Czech Technical University, Prague, 1982.

[2.6]   Greunen, I. van: Nonlinear, Geometric, Material and Time Dependent Analysis of Reinforced and Prestressed Concrete Slabs and Panels. Report No. UC SESM 79-3, University of California, Berkeley, 1979

[2.7]   Křístek, V. – Šmerda, Z.: Simplified Calculation of the Relaxation of Stress Respecting the Delayed Elasticity. Proceedings of the Conference on Fundamental Research on Creep and Shrinkage of Concrete. Wittmann, F. H. (Ed.), Martinus Nijhoff Publ., The Hague–Boston–London, 1982.

[2.8]   Křístek, V. – Máca, J. – Šrůma, V. – Vítek, J. L.: Simplified Calculation of Stress Relaxation in Concrete. Prestressed Concrete in Czechoslovakia. Inženýrské stavby 5–6, 1982.

[2.9]   Trost, H.: Auswirkungen des Superpositionsprinzip auf Kriech- und Relaxationsprobleme bei Beton und Spannbeton. Beton- und Stahlbetonbau, 11/1967.

[2.10]   Bažant, Z. P.: Prediction of Concrete Creep Effects Using Age-Adjusted Effective Modulus Method. ACI Journal, April 1972, pp. 212 – 217.

[2.11]   Bažant, Z. P. – Kim Sang-Sik: Approximate Relaxation Function for Concrete. Journal of the Structural Division, ASCE, Proc. Paper 15083, December 1979, pp. 2695–2705.

[2.12]   Rüsch, H. – Jungwirth, D.: Stahlbeton. Berücksichtigung der Einflüsse von Kriechen und Schwinden auf das Verhalten des Tragwerke. Düsseldorf, 1976.

[2.13] Buseman, R.: Kriechberechnung von Verbundträgern unter Benutzung von zwei Kriech-fasern. Der Bauingenieur No. 11, Berlin, 1950.

[2.14] Leonhardt, F.: Spannbeton für die Praxis. Berlin, 1973.

[2.15] Recommendations Internationales CEB–FIP pour le Calcul et l'Exécution des Ouvrages en Béton. Paris, 1970.

[2.16] Règles Unifiées Communes aux Différent Types d'Ouvrages et de Matériaux Vol. I, Code Modèle CEB–FIP pour les Structures en Béton Vol. II, CEB–FIP, April 1978.

[2.17] Ulickiy, I. I.: Stiffness of reinforced concrete members subjected to bending. (In Russian.) Akademia stroitelstva i arkhitektury USSR, Kiev, 1963.

[2.18] ACI Committee 209, Subcommittee 2, Prediction of Creep, Shrinkage and Temperature Effects in Concrete Structures. Detroit, 1971.

[2.19] Bažant, Z. P.–Panula, L.: Practical Prediction of Creep and Shrinkage of Concrete. Matériaux et Constructions (RILEM, Paris), Parts I and II, No. 69, 1978, Parts V and VI, No. 72, 1979.

[2.20] Ulickiy, I. I.: Theory and analysis of reinforced concrete frame structures with consideration of long-term processes. (In Russian.) Kiev, 1967.

[2.21] Bažant, Z. P.–Chern, J. C.: Triple Power Law for Concrete Creep, Journal of Engineering Mechanics, ASCE, Proc. Paper 19397, Vol. 111, No. 1, January 1985, pp. 63–83.

[2.22] Bažant, Z. P.: Constitutive Relations, RILEM TC–69 Mathematical Modelling of Creep and Shrinkage of Concrete, State-of-Art Report, 1986.

[2.23] Bažant, Z. P.: Mathematical Model for Creep and Thermal Shrinkage at High Temperature, Nuclear Engineering and Design 76 (1983), pp. 183–191.

[2.24] Bažant, Z. P.–Chern, J. C.: Bayesian Statistical Prediction of Concrete Creep and Shrinkage, ACI Journal, July–August 1984, pp. 319–329.

[2.25] Bažant, Z. P.–Chern, J. C.: Concrete Creep at Variable Humidity, Report No. 84–5/665c, Northwestern University.

[2.26] Bažant, Z. P.–Wang, T. S.: Spectral Finite Element Analysis of Random Shrinkage in Concrete, Journal of Structural Engineering, ASCE, Vol. 110, No. 9, September 1984, pp. 2196–2211.

[2.27] Bažant, Z. P.–Chern, J. C.: Strain Softening with Creep and Exponential Algorithm, Journal of Engineering Mechanics, ASCE, Vol. 111, No. 3, March 1985, pp. 391–413.

[2.28] Kabir, A. F.: Nonlinear Analysis of Reinforced Concrete Panels, Slabs and Shells for Time–dependent Effects. PhD Dissertation, University of California, Berkeley, Report No. UC–SESM 76–6.

[2.29] Scordelis, A. C.: Computer Models for Nonlinear Analysis of Reinforced and Prestressed Concrete Structures, PCI Journal, November–December 1984.

[2.30] Bažant, Z. P.: Theory of Creep and Shrinkage in Concrete Structures – A Précis of Recent Developments. Mechanics Today, Vol. 2, Pergamon Press, 1975.

Chapter 3 – References

[3.1] Křístek, V.–Vítek, J. L.: Method of Effective Time in Concrete Creep Analysis. Proc. of the RILEM Conf., Budapest, 1984.

[3.2] Bažant, Z. P.–Kim Sang-Sik: Approximate Relaxation Function for Concrete. Journal of the Structural Division, ASCE, Proc. Paper 15083, December 1979, pp. 2695–2705.

[3.3] Bažant, Z. P.–Wittmann, F. H.: Creep and Shrinkage in Concrete Structures. J. Wiley and Sons, Chichester–New York–Brisbane–Toronto–Singapur, 1982.

[3.4] Glanville, W. A.: The Creep or Flow of Concrete under Load. Build. Research, Techn. paper No. 12, 1930.

[3.5]   Dischinger, F.: Elastische und plastische Verformungen der Eisenbetontragwerke und insbesondern Bogenbrücken. Bauing. 1939.

[3.6]   Trost, H.: Auswirkungen des Superpositionsprinzips auf Kriech- und Relaxationsprobleme bei Beton- und Spannbeton. Beton und Stahlbetonbau 11/1967.

[3.7]   Ulickiy, I. I.: Theory and analysis of reinforced concrete frame structures with consideration of long-term processes. (In Russian.) Kiev, 1967.

Chapter 4 — References

[4.1]   Bažant, Z. P.: Theory of Creep and Shrinkage in Concrete Structures: A Précis of Recent Developments. Mechanics Today, Vol. 2, Pergamon Press, 1975.

[4.2]   Bažant, Z. P.: Creep of concrete in the analyses of structures.(In Czech.) SNTL Prague, 1966.

[4.3]   Bažant, Z. P. — Wittmann, F. H.: Creep and Shrinkage in Concrete Structures. J. Wiley and Sons, Chichester — New York — Brisbane — Toronto — Singapur, 1982.

[4.4]   Bažant, Z. P. — Liu, K. L.: Random Creep and Shrinkage in Structures: Sampling, Journal of Structural Engineering, ASCE, Proc. Paper 19725, Vol. 111, No. 5, May 1985, pp. 1113 — 1134.

[4.5]   Křístek, V. — Vítek, J. L.: Calculating for Concrete Creep — An Approximate Method of Effective Time. Building Research and Practice, Vol. 14, No. 2, March — April 1987, pp. 102 — 108.

Chapter 5 — References

[5.1]   Glanville, W. H. — Thomas, F. G.: Studies in Reinforced Concrete IV. Further Investigations of the Creep or Flow of Concrete under Load. Build. Research, Techn. Paper 21, 1939, London.

[5.2]   Evans, R. H.: Some New Facts Concerning Creep in Concrete. Concrete and Constructional Engineering 37, 1942, 12.

[5.3]   Ulickiy, I. I.: Theory and analysis of reinforced concrete structures with consideration of long-term processes. (In Russian.) Kiev, Budivelnik, 1967.

[5.4]   Nemirovskiy, Ya. M.: Investigation of stressed and deformed state of reinforced concrete members with consideration of the work of tensioned concrete over cracks and revision of the theory of the analysis of deformations and discovery of cracks on this basis. Strength and rigidity of reinforced concrete structures. (Ed. Gvozdyev.) (In Russian.) NIIZHB, Moscow, 1968.

[5.5]   Vasilev, B. F.: Analysis of reinforced concrete structures for strength, deformations, development and discovery of cracks. (In Russian.) Moscow, 1965.

[5.6]   Ulickiy, I. I. et al.: Stiffness of reinforced concrete members subjected to bending. (In Russian.) Kiev, Akademia stroitelstva i arkhitektury USSR, Budivelnik, 1963.

[5.7]   Nemirovskiy, Ya. M. — Kochetkov, O. I.: Effect of work of compressed and tensioned zones of reinforced concrete on deformations of simple members subjected to bending after development of cracks in them. Peculiarities of deformations in concrete and reinforced concrete. (Ed. Gvozdyev.)(In Russian.) NIIZHB, Moscow, 1969.

[5.8]   Neville, A. M.: Creep of Concrete. Plain, Reinforced and Prestressed. North Holland Publ. Comp., Amsterdam, 1970.

[5.9]   Recommandations Internationales CEB–FIP pour le Calcul et l'Exécution des Ouvrages en Béton. Paris, 1970.

[5.10] Branson, D. E.: Instantaneous and Time Dependent Deflections of Simple and Continuous Reinforced Concrete Beams. Part I. Alabama Highway, Research Report No. 7, August 1963.

[5.11] CEB Bulletin d'Information 1967, No. 62, No. 63.

[5.12] CEB Bulletin d'Information 1972, No. 79.

[5.13] CEB Bulletin d'Information 1971, No. 77.

291

Chapter 6 — References

[6.1]   Instruction for the design of reinforced concrete thin-walled three-dimensional structures. (In Russian.) Stroyizdat, Moscow, 1961.

292

Index